SUGGESTIONS

Also by James L. Weber

Genomes and the Gene Pool (2026)

SUGGESTIONS

for Long Term Goals for Humanity

JAMES L. WEBER

TEN GENERATIONS PRESS

Published by Ten Generations Press, Marshfield, Wisconsin

www.tengenerationspress.org

Suggestions for Long Term Goals for Humanity

Publisher's Cataloging-in-Publication
(Provided by Cassidy Cataloguing Services, Inc.)

Names: Weber, James L., author.
Title: Suggestions for long term goals for humanity / James L. Weber.
Description: Marshfield, Wisconsin : Ten Generations Press, [2026] | Includes bibliographical references.
Identifiers: LCCN: 2026918234 | ISBN: 9798995825357 (hardcover) | 9798995825333 (paperback) | 9798995825340 (ebook)
Subjects: LCSH: Progress. | Social problems. | Social change. | Humanity. | Cooperation. | Civilization. | BISAC: SOCIAL SCIENCE / Future Studies. | PHILOSOPHY / Social. | POLITICAL SCIENCE / International Relations / Treaties.
Classification: LCC: HM891 .W43 2026 | DDC: 303.44--dc23

Dedicated to the legions who have
labored in anonymity

Table of Contents

List of Tables and Figure

Preface

THERE ARE VERY FEW COMPLETELY novel ideas for improvement of human societies. Billions of people have pondered important human questions and problems over millennia. I therefore don't claim originality for any of the ideas in this slim book. I do, however, claim novelty for my particular combination of ideas.

Because at various times throughout history, so many people have expressed the ideas presented in this book, it is impossible for me to properly credit each one. I apologize in advance to all those, past and present, who have not been properly credited.

The references listed at the end of the book are organized by chapter and cited by number in the text. I did not attempt to be comprehensive, but rather included references that were relatively recent, broad, authoritative, and not too technical. In many cases, the references include a PubMed identification number. This number can be used to easily access the article through the US National Library of Medicine's PubMed website.

Although I'm not a great writer, I think I have over the years learned to write reasonably clearly and concisely. While not entertaining, this book has the virtue of being short. Readers can get through it in a relatively small amount of time. Artificial intelligence was not used to write any part of this book.

All income from sale of this book will be donated to charity.

Like all of us, I benefit from millions who, through enormous effort, have contributed ideas, knowledge, and beliefs to human societies. I especially thank all of my teachers, mentors, and colleagues who helped me learn and grow, and tolerated my many foolish mistakes.

Chapter 1

Long Term Goals for Humanity

PEOPLE EXPEND ENORMOUS ENERGY PURSUING day-to-day activities. With gusto we work, play, and live, but our societies make only very slow progress. Major problems like warfare, inequality, and environmental degradation continue to fester decade after decade, century after century. Why, despite our enormous collective capability, do our societies advance so slowly? In large part, I think it's because we lack direction. The human ship is rudderless. If we could increase, even by a small amount, the fraction of our collective time and energy devoted to achieving worthwhile long term goals, humanity would leap ahead.

A goal is an objective toward which our efforts are directed. People set goals because they focus our efforts and straighten the path toward our objectives.

Our species needs goals for the same reasons that individuals and organizations need goals.

Certainly, many individuals are working hard today, even devoting their entire lives, to improving the human condition. These heroines and heroes deserve great praise. But even these well-intentioned, honorable people are pulling in different directions without unified effort. Long term goals will help channel these efforts.

Formulating long term goals will force us to carefully evaluate where we have been, where we are, and where we should be going. Many recent technological advancements have raced ahead of our understanding. Setting long term goals will help us cope with these rapid, unsettling changes.

Politicians rarely look further ahead than the next election. Long term goals will help them to at least occasionally consider how their actions will affect humanity's more distant future. Political leaders are also faced with a multitude of demands for limited resources. Long term goals will help them prioritize those demands.

Why long term rather than short term goals? Because short term actions and goals are often inefficient and sometimes even wasteful. Consider, for example, all the billions of dollars that have been spent over the centuries treating diarrheal diseases like cholera, when providing people with clean, safe drinking water would largely eliminate the need for these treatments. Accomplishing worthwhile long term goals will get at the roots of problems, rather than continually applying short term fixes. Long term goals are also required because some human problems are so difficult that it will take many decades or even centuries to achieve solutions.

Goals also influence personal conduct. If a person's goal is to own a large oceanfront home, then they will pursue the wealth necessary to purchase such a home. If their goal is to help their students learn as much as possible, then they will devote their time and energy to teaching. Setting long term goals for humanity will encourage and motivate many people to work toward accomplishment of these goals.

Finally, long term goals will help us appreciate trends and time. Even small rates of change, for better or worse, continued for centuries can have enormous impacts on societies. Several of the goals described in this book deal with trends that are currently heading in the wrong direction. Species are diminishing in the wild; totalitarianism is spreading; deleterious DNA variants are accumulating in the gene pool. I don't wish to be gloomy, but if these trends continue decade after decade, century after century, humanity is in trouble.

- - -

What types of goals are people setting today? Many individuals set personal goals, some of which endure over lifetimes. Public and private nongovernmental organizations of all sorts also often set goals. Even governments occasionally set goals. None of these goals, however, are long term, and none encompass all of humanity.

Some specific pan-humanity goals have arisen from international treaties. Probably the best-known current example is the goal of limiting global warming to 1.5°C from the Paris Agreement on climate change adopted in 2015 (1). Another example is the goal of eliminating the production of chemicals that destroy atmospheric ozone from the Montreal Protocol finalized in 1987 (2). These treaty goals are shorter term and more focused than the goals presented in this book.

To my knowledge, the closest humanity has come to setting broad, planet-wide goals are the United Nations (UN) Development Goals (3). The first set, the eight Millennium Development Goals, were adopted in 2000 to be achieved by 2015. The second set, 17 Sustainable Development Goals, were adopted in 2015 to be accomplished by 2030. Each of the Sustainable Development Goals has targets (169 in all), and importantly, each of the targets is accompanied by indicators to measure progress toward the goals.

I think the UN Development Goals are a wonderful advance for humanity. They are the first universally approved set of goals for our species. There is some overlap between the UN goals and the goals described in this book, but the UN goals differ from my suggested goals in that they are much shorter term and are focused on relatively narrow topics like education. Because the UN encompasses all nations, its goals also could not cover contentious topics such as human rights. In addition, the UN goals are exceptionally idealistic and ambitious. Probably none of the 17 major goals, and few of the 169 targets, will be achieved by 2030 (4).

- - -

My list of seven interrelated long term goals is shown in the following table. For this book, I define "long term" as goals that are intended to be achieved in 300 years (about 10 generations). My suggested goals are planet-wide; they are not restricted to any one nation or region. I devote a separate chapter in the book to each goal. In Chapter 9, I offer some ideas on how to establish and accomplish the goals.

Table 1
My list of suggested long term goals for humanity

Goal	Chapter
Continue the Human Species	2
Continue Other Species	3
Establish a Democratic World Government	4
Manage the Human Gene Pool	5
Grow Knowledge	6
Temper Greed	7
Preserve Knowledge	8

For each goal, I attempt to clearly describe *what* the goal involves, explain *why* the goal is important, and offer some ideas about *how* to achieve the goal. Consistent with the UN Development Goals, I also provide measurable targets. I think that nebulous goals without clear, measurable targets, such as "expand human rights," are not very useful.

During my lifetime, I have prepared thousands of sets of goals both for myself and for various teams, departments, and organizations. These experiences have taught me that the best goals are those that are difficult to achieve, yet can still be accomplished. Goals that are easy to achieve don't sufficiently motivate people. Goals that are overly ambitious and impossible to achieve are frustrating and demoralizing. It's difficult to formulate goals with the correct level of ambition, but at least we can try to avoid goals that are far too easy or wildly impractical.

To accomplish the long term goals, humanity will certainly need capable individuals who will dedicate much of their lives to these tasks. But I don't think that important human goals will be achieved by relatively few individuals working in isolation, nor even by small groups, but rather by the combined actions of millions. Therefore, in each chapter I also provide lists of pragmatic steps that all people can take to help accomplish the goals. People living in totalitarian states have fewer opportunities to contribute, but can still make a difference.

Detailed plans for achievement of the goals will certainly be required. Subgoals and milestones will also be needed. While I do offer a few ideas for achieving the goals, I've made no attempt to construct detailed plans.

The problems addressed by my list of goals are certainly not the only problems that humanity needs to solve. Among many possible long term goals, I selected for my list those that I think will have the greatest impact. Achievement of the goals listed in this book will facilitate much additional progress.

- - -

Critics of this book will say that it is relatively easy to devise lists of goals, but vastly more difficult to achieve them. I agree completely. To accomplish the goals, many new approaches will be needed, and many obstacles will need to be overcome. I spent about 20 years leading a small private company. I dealt daily with the seemingly endless vexing problems that always arise. I understand perfectly well how difficult it is to accomplish worthwhile goals.

Yet I still claim that long term goals are essential. Beyond focusing our energies, they also inspire. Large fractions of people around the globe today are pessimistic about the future. Falling birth rates may be a symptom of this pessimism. Setting and working toward long term goals will help dispel this widespread gloom. People will always have self-interest, but we also have a deep and real need to contribute to something grand. Consider the construction throughout history of exquisite monuments, temples, churches, and mosques. Worthy long term goals will give us hope.

Chapter 2

Continue the Human Species

WHAT

This goal means that our species should not become extinct over the next 300 years. It also means that we should develop knowledge, practices, and policies that will allow humanity to continue much further into the future, hopefully for millions of years.

To appreciate this goal, it's useful to have at least a rudimentary knowledge of human evolutionary history (1–5). Very briefly, our planet is old, about 4.5 billion years old. Simple living organisms are also very old. The first known unicellular organisms date to around 3.7 billion years ago. Gradually, over periods of many millions of years, the first relatively simple organisms evolved into more complex species. Some key dates in human evolutionary history are shown in Table 2. Since scientists are still learning a great deal about the history of life on our planet, I expect that the dates shown in Table 2 will be revised. Still, they offer important perspective.

Table 2
Some key dates in human evolutionary history

Event	Years Ago
Formation of Earth	4.5 billion
First living organisms	3.7 billion
First animals	600 million
Earliest land plants	470 million
First mammals	230 million
Demise of the dinosaurs	66 million
Last common human-chimp ancestor	7 million
Appearance of Neanderthals	600,000
First modern humans	200,000
Start of emigration of modern humans throughout the world	60,000
Extinction of Neanderthals	40,000
Advent of agriculture	12,000

From Table 2, it can be seen that humans are a very young species. We just recently made our appearance on Earth's stage. Anatomically modern humans have been around only about 200,000 years or 0.00005 (five–one hundred thousandths) of the time that there has been life on our planet. In comparison, although the species are different from those that exist today, animals that we would all recognize as fish have been around for about 530 million years, dragonflies for 320 million years, and snakes for 130 million years. Snakes are therefore 650 times older than people.

Today, our species dominates the planet. It is difficult to imagine that we could become extinct. But we need to keep in mind that many species that were once highly successful are now extinct. Consider the large and powerful dinosaurs. They were at the apex of ecosystems for over 150 million years. Yet they were wiped out in an evolutionary eyeblink by a 10 kilometer diameter asteroid that smacked into the Yucatán Peninsula in Mexico about 66 million years ago (6–7).

Or consider passenger pigeons, a species of bird from my part of the world (central North America) (8). When large numbers of Europeans arrived in central North America in the 19th century, passenger pigeons were the most abundant bird in North America and were estimated to number about three billion (comparable to the current human population of eight billion). Some of the flocks were so huge that they took hours to pass overhead. John James Audubon, iconic American ornithologist, said in 1813 of passenger pigeons: "The air was literally filled with Pigeons; the light of noon-day was obscured as by an eclipse."

Hunters killed vast numbers of birds. In 1882, for example, hunters in the US state of Wisconsin shipped 75 barrels of birds to cities each day, with each barrel containing 500 birds. The last of the wild birds is thought to have been killed in 1901, and the last bird in captivity died in 1914. So, passenger pigeons went from a population of three billion to zero in just a few decades.

Or consider our close cousins the Neanderthals (and the related Denisovans). When modern humans migrated out of Africa about 60,000 years ago, Neanderthals were a successful species occupying much of Europe and Asia. They had been living there for hundreds of thousands of years. Yet within a relatively short time they were gone.

When I was a young boy, few things in life seemed as strong to me as trees. They were so tall and hard. Yet with our present technology, we can easily destroy entire forests. Unlike in science fiction movies, if a technologically advanced species ever did travel to Earth and decide to exterminate humans, it would be trivially easy for them. As easy as it is for us to destroy a forest.

Toby Ord, the author of *The Precipice* (9), estimated the chance that humanity will experience a species-threatening event in the next century at 1 in 6 (18%). This is a startlingly high number that should serve as a wake-up call for us all.

- - -

A number of scholars have attempted to catalog all known threats to our species (9–12). Among these threats, I outline here only five: artificial intelligence, asteroids, pathogens, supervolcanoes, and the gene pool. Although the risk to human extinction from any one of these threats in any year is low, collectively, over time, the risks become substantial. If we want humanity to survive for millions of years, then we must study and prepare for such threats.

Perhaps the greatest current threat is artificial intelligence (AI). Computer programs have become so powerful that they can mimic many human cognitive abilities, including playing complicated games, taking tests, writing articles, and holding conversations. These programs are improving so rapidly that many computer scientists think they will soon surpass humans in all cognitive abilities and become superintelligent (13–14). Experts debate whether these programs will become conscious or sentient, but regardless, they will become exceptionally powerful and will greatly affect human societies.

Some project that these superintelligent programs will usher in a new era of progress and prosperity for humanity (15). Others think that these programs will be so far beyond us in intelligence that they will be able to easily control humans through tricks, bribes, and threats. It is even possible that the programs will someday decide that they are better off without humans and destroy us (16). In 2023, hundreds of tech leaders signed on to the following statement (17):

> Mitigating the risk of extinction from AI should be a global priority alongside other societal-scale risks such as pandemics and nuclear war.

Over perhaps the next two decades, AI could become the greatest threat humanity has ever known.

- - -

As mentioned above, an asteroid struck Earth about 66 million years ago and extinguished the dinosaurs and many other species. Those who doubt the immense power of such a collision should watch the videos made through our telescopes of Comet Shoemaker–Levy 9 colliding with the planet Jupiter in 1994. The impacts released energy equivalent to six million megatons of TNT, or about 600 times the energy of the world's total nuclear weapon arsenal (18).

The asteroid/comet threat has been recognized for decades (9, 19–20). Astronomers have at the time of writing cataloged the orbits of nearly all of the larger threatening asteroids (greater than about 500 meters in diameter) and many of the smaller ones (greater than 150 meters in diameter) (21). However, not all dangerous objects in our solar system have been cataloged. Relatively few long-period comets have been identified, and three interstellar objects that transited our solar system were recently discovered (22–23).

Approaches to defeat an asteroid or comet strike have begun to be tested. The Double Asteroid Redirection Test (DART) space mission involved the deliberate collision of a space probe with a distant asteroid to determine whether the asteroid could be deflected (24–25). The probe successfully shifted the orbit of the asteroid. But this was just a first, crude attempt. A great deal of work remains.

- - -

When I was a biochemistry graduate student in the late 1970s, I read a novel about a man-made plague that nearly extinguished humanity (26). While I enjoyed the novel, I didn't think at that time that our technology was sufficiently advanced to create a new pathogen. Fast forward 45 years. I no longer think that our technology is insufficient. In fact, researchers, in attempts to better understand pathogens, have already accomplished the feat (see, for example, 27–28). Such work today still requires elaborate, well-funded labs with special expertise, but as technology progresses, it will become possible for smaller, less sophisticated labs to produce deadly pathogens.

As I write, humanity is still recovering from the worst pandemic in a century, the COVID-19 coronavirus pandemic. The pandemic, which killed about 20 million people (29–30), sickened at least hundreds of millions, and caused massive economic disruption, reacquainted us all with the danger of pathogens. If a more deadly pathogen arises in future, either natural or man-made (31), or even worse, if two or more such pathogens begin spreading at the same time, then humanity could indeed be threatened.

- - -

Large volcanic eruptions, often called supervolcanoes, can spew enough dust into our planet's atmosphere to block sunlight and significantly reduce plant growth (32–34). Scientists have gathered convincing evidence that such eruptions have occurred in our planet's past. With less food, millions would starve. If a large enough single eruption or series of eruptions occurred close in time to one another, our species could become threatened. Scientists have also projected that a worldwide nuclear war with concomitant massive fires could have the same sunlight-blocking effect as supervolcanoes (35–36).

- - -

Through development of our modern societies, we have unintentionally shifted the trajectory of human evolution. In particular, we have substantially relaxed natural selection. This change and other recent changes in evolutionary forces are already altering the human gene pool (the collective genomes of humanity). This threat is not yet widely recognized but, in the long term, has the potential to cause great harm. *Homo sapiens* will likely survive this threat, but if our descendants can no longer sustain civilization, humanity as we know it today could become extinct. The gene pool threat is discussed briefly in Chapter 5 in this book, and in much more detail in Part II of another book I have written (*Genomes and the Gene Pool*) (37).

WHY

Although complex living organisms have been present on our planet for hundreds of millions of years, intelligent life capable of creating civilizations evolved only once. I conclude from this fact that evolution of species as intelligent as ourselves is a rare event, perhaps exceptionally rare.

Is there life, and especially intelligent life, in the universe outside of Earth? The universe is so immense that it is certainly possible. Our sun is just one of at least a hundred billion stars in our galaxy, and our galaxy, in turn, is just one of hundreds of billions of galaxies in the universe. If intelligent life is common in the universe, then from the perspective of the universe, humanity's survival may not be that important. But if life, and especially intelligent life, is rare in the universe or even conceivably absent from everywhere but Earth, then our planet and the species on it are gems of inestimable price.

Despite all our many mistakes and problems, I still have pride in our species. In addition to all the horrible crimes that humans have committed, we also can point to many wonderful achievements. Imagine what our descendants might do if we can continue for thousands and hopefully millions of years, and if our pace of scientific and technological advancement continues. We could perhaps travel to other stars or seed life on billions of planets. A virtually unlimited number of marvelous projects and achievements are possible, *but* our species needs to survive to get there.

Finally, I hate failure, and I think most other people do as well. It was estimated in 2002 that 81 billion people once lived or were currently alive (38). Those billions put forth enormous effort and made astounding sacrifices to get us where we are today. Extinction of our species would be a mockery and negation of all their efforts and a monumental failure. I and many others (39) believe that we have a duty to ensure that our children and their descendants survive.

HOW

There are several steps we can take to mitigate existential threats to our species. One important step will be to greatly expand study of possible threats, particularly at universities and other nonprofits. Such study will involve identifying, analyzing, and ranking threats as well as devising preventive measures. There may well be threats that have not yet been identified. Although there are individuals at many institutions around the world studying disasters, there are very few focused on existential risks.

World government and governments of at least large nations need to establish departments with the mission of moderating and, where possible, preventing catastrophic and existential events. The use of taxpayer money to fund such government departments can be justified by our need to deal with relatively frequent catastrophes, such as earthquakes and epidemics, which, although they do not place human survival at risk, still result in the deaths of hundreds or thousands or millions. Clearly, there is substantial overlap between events that kill large numbers and those that might extinguish our species. We can leverage our desire to allay less devastating but more frequent events to fund efforts to reduce existential threats.

We also need to take steps toward establishing at least one human refuge. This would be an independent, self-sufficient colony of people that could continue our species even if the rest of humanity was eliminated. Such refuges could conceivably be underground on Earth, but underground refuges probably would eventually be susceptible to the same threats as above-ground populations. Better would be to establish these refuges in space.

I hope I'm wrong, but I doubt that truly independent space refuges can be achieved in 300 years. A great deal of new technology and capability will need to be developed. We should, however, be able to make significant progress toward this aim over the next three centuries. My opinion is that the *top* priority of space programs worldwide should be the continuation of humanity. I am a strong proponent of scientific research (Chapter 6), including research in space, but I still feel that the majority of space program resources should be devoted not to research, but rather to threat deterrence and establishment of refuges.

Especially when people are waiting or find themselves in an unpleasant situation, even a single day seems like an eternity. Yet, one day or even one year is just a minuscule fraction of the time that our planet has existed or that life has been on the planet, or even that anatomically modern humans have been around. For humans to survive into the distant future, we need to expand our time horizons from mere years to centuries and millennia. We need to improve our understanding and appreciation of time.

TARGETS

- Increase the number of university departments and other nonprofit institutions worldwide that are devoted to the study and reduction of existential threats to at least 200.

- Increase the number of government departments worldwide devoted to catastrophic risk planning and mitigation to at least 50. This would include a department at the world government level.
- Increase annual worldwide funding for efforts to study and defeat threats to our species by a factor of at least 1,000. This should not be difficult given that current spending is so low.
- Devote at least 50% of annual space budgets to threat deterrence (primarily asteroids and comets) and to creation of independent space refuges.
- Establish term limits for *all* elected officials so that our leaders, at least occasionally, look further ahead than the next election.

ACTIONS WE CAN ALL TAKE

- Advocate for increased study of threats to our species.
- Donate money to support such studies.
- Support efforts to deter terrorism, especially bioterrorism.
- Always treat all machines, including computers, as tools; *never* accept them as masters.

Chapter 3

Continue Other Species

WHAT

This long term goal is to preserve as many nonhuman species as possible. This goal covers *all* species, not just charismatic species like pandas and polar bears.

Scientists who have carefully studied the fossil record have identified several major extinction events that occurred in our planet's distant past (1–2). Many theories have been advanced to explain these events, but scientists know with certainty the cause of only one: the collision of a large asteroid with the earth about 66 million years ago. Many believe that another major extinction event is currently underway, this one caused by human activity.

Although in some cases there may have been contributing factors, it is known that our ancestors caused the extinction of many species (3). To cite just a few examples, when the ancestors of Native Americans migrated into North America about 16,000 years ago, they extinguished a number of large mammals, including giant ground sloths, mammoths, and saber-toothed cats. The Polynesian Māori terminated the moas, large ostrich-like birds, when they entered New Zealand in about the year 1350. And as described in the previous chapter, European immigrants to central North America wiped out the passenger pigeon in the 19th and early 20th centuries.

Figure 1
Growth in human world population

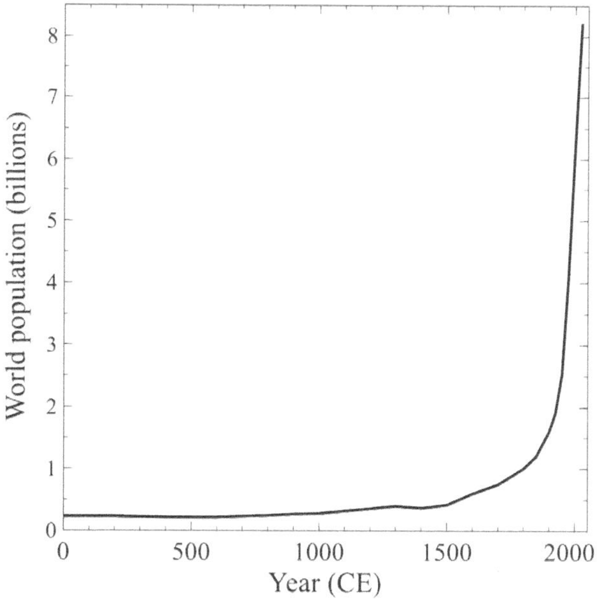

Estimates of world population by year in the Common Era (CE). Adapted from references 4–6.

Human-caused extinctions continue today. The human population has exploded since the year 1800, going from about one to eight billion people (Figure 1)

(4–6). Our greatly increased population has necessitated the conversion of vast tracts of our planet from the natural or near natural state into farmland, cities, and other areas of intense human modification. It is estimated that up to 75% of Earth's land area and 66% of ocean area have been significantly altered by human activity (7–8).

The conversion of a large fraction of our planet's surface to human usage has relegated many organisms to greatly reduced and fragmented habitats. Not surprisingly, the reduction in habitat has resulted in great declines in population numbers (9–12). Curators of the Living Planet Index report a 69% decline in population numbers of monitored species between 1970 and 2018 (13–14). The Index primarily covers vertebrates. There have also been dramatic declines in insects (15–17), mollusks (18), plants (19), and many other species.

It is difficult to monitor the populations of many wild species, and most of the estimates of population changes have come from people who are devoted to wildlife preservation. It is therefore possible that some of the declines are exaggerated (20). But there is no doubt that *far* more species are declining than are gaining numbers, and that many of the declines are dramatic (21). Even casual observers like myself who have lived several decades can testify firsthand to the decline in many species over our lifetimes.

In some cases, the declines in species numbers have progressed all the way to extinction. Extinctions have always occurred, but the pace of prehuman extinctions was far slower than the current rate of extinctions. We have increased the rate of extinction almost certainly by a factor of at least 10, and possibly by a factor as large as 1,000 (8, 22–23). Substantial fractions of vertebrates and other species are threatened (Table 3). Extrapolation of current declines in species numbers into coming decades portends that huge numbers of species could be lost.

Table 3
Fractions of selected groups of species threatened with extinction (24)

Group	Number of Species	Fraction of Species Threatened
Mammals	5,000	27%
Birds	10,000	13%
Amphibians	6,000	41%
Reptiles	9,000	21%

I feel that we should largely forgive our ancestors for past extinctions, especially those that took place hundreds or thousands of years ago. Our ancestors didn't possess the data and technology that we have today. They may have had some inkling of what they were doing, but basically, they were just trying to survive. The current situation, however, is completely different. We know with certainty that we are causing the extinction of massive numbers of species. We know that we can preserve other species and still maintain our modern societies.

WHY

The reasons for preserving other species can be split into two groups: moral and pragmatic. One moral argument was introduced in the previous chapter. If life in the universe outside of Earth is rare or even absent, then life on Earth is extraordinarily precious. Imagine that intelligent aliens are monitoring our actions. What, from the alien viewpoint, would be humanity's greatest crime? I think extinguishing other species would likely be at the top of the list. Killing each other is foolish and horrible, but at least we are doing this to ourselves. Other species, however, are defenseless against human intelligence and technology.

Or, for those who are religious and believe that God (or gods) created the earth and all living things, how could it be imagined that God is happy about our deliberate destruction of his creation. Truly impressive theological gymnastics are necessary to try to justify our actions toward other species.

As a final moral argument, our planet is expected to exist for at least another billion years before being fried by an expanding sun. Hopefully, life will continue to thrive on Earth all through this very long time. Evolution of intelligence at the human level has so far happened only once on our planet and is therefore a rare event. But over a billion years, it may be possible for other species to evolve human-level intelligence. I value all living things, but I value foremost those species with the highest levels of intelligence, such as dolphins and chimpanzees. These species are probably the most likely to someday evolve human-level intelligence. Even if humans become extinct, perhaps another intelligent species will take our place.

- - -

On the pragmatic side, other species have enormous financial value. In a landmark paper, a group of economists in 1997 calculated that the financial value of other species was greater than the total world gross national product (25). Others have since confirmed the huge monetary value of nature (26–27).

There are many ways in which we benefit from other species (8, 13, 27–28). Some of the major ways are listed in Table 4 and briefly described below.

Table 4
Some of the ways in which humans gain from wild species

- Provision of food
- Purification of water and air
- Removal of carbon dioxide from the planet's atmosphere
- Enhancement of rainfall on crop and pastureland
- Global recycling of elements
- Provision of lumber, paper, and fuel
- Prevention of erosion and flooding
- Provision of drugs, proteins, and other valuable biomolecules
- Enhancement of recreation, tourism, and property value

Millions of people around the world still depend upon uncultivated species for food, especially wild fish and mollusks (clams, oysters and the like). Wild nuts, fruits, and mushrooms are also consumed in large quantities. Bees, other insects, and some birds pollinate many of our agricultural crops. Forests provide us with wood for lumber, paper, and fuel. Wild relatives of cultivated crops are a valuable resource for plant breeding to improve the quality and yield of agricultural products (29).

Organisms clean our water and air by removing toxic compounds. Plants and other photosynthetic organisms ameliorate global warming by absorbing carbon dioxide from the atmosphere.

Forests enhance rainfall. Reduction in the size of the Amazon rainforest is predicted to increase temperatures and decrease rainfall in South America, with concomitant reductions in agricultural yields (13, 19, 27, 30). Plants buffer erosion of soils from our cropland, hillsides, and shorelines (13).

Organisms play major roles in the global cycling of essential chemical elements such as carbon, oxygen, sulfur, nitrogen, and phosphorus (see, for example, 13, 28). Much is still being learned about how animate and inanimate processes are involved in the global cycling of elements, but disrupting these cycles by destroying essential organisms could be devastating to human welfare.

Many of our drugs and other useful biomolecules were initially identified in other species. Here are just a few examples. The antibiotic penicillin comes from a mold. The antimalarial drug artemisinin comes from sweet wormwood, an herb used in traditional Chinese medicine. The DNA-synthesizing enzyme Taq polymerase comes from the bacterium *Thermus aquaticus*, which lives in hot springs. Taq polymerase is used today in thousands of different clinical tests, including those used to detect the viruses that infect our bodies. There are undoubtedly millions of other useful biomolecules that remain to be discovered. Some of these will come from threatened exotic species living in very specialized habitats. When these organisms are gone, we lose their valuable biomolecules.

People place great financial value on the aesthetics of nature (31). For instance, the value of property alongside rivers, lakes, and oceans is much higher than equivalent property away from water. But when the water becomes so polluted that fish die, algae bloom, and species vanish, then property value is greatly diminished.

HOW

Unlike the threats to our own species, extinction of other species has received wide attention from biologists and from the press. Many species are being monitored. Thousands of people worldwide are actively working to preserve species.

These experts have developed long lists of actions that we can all take to continue other species (13). A few of these actions are presented at the end of this chapter. I think, however, that two steps are especially important. One is to reserve substantial fractions of the surface of our planet for other species. The other is to slowly reduce human population numbers so that we ease the stress on ecosystems.

Species are dependent upon specific habitats. To mention just a few of the many thousands of known examples, orangutans need jungle on Borneo and Sumatra, manatees in the Gulf of Mexico need coastal rivers, and the small Karner blue butterfly needs a species of lupine that grows only in unforested areas of North America. When the Southeast Asian jungles are cleared to make way for palm oil plantations, when the river seagrass upon which the manatees feed dies due to pollution, and when the Karner blue prairies are plowed under to grow maize and potatoes, then these species are at risk for extinction. Even when small patches of habitat are preserved, species become threatened. They become genetically weakened through inbreeding and are more susceptible to chance events like storms or epidemics. To quote Charles Darwin, "Rarity is the precursor to extinction."

It's also important to keep in mind that individual species do not live in isolation, but rather within ecosystems. When one species declines or becomes extinct, the other species within the ecosystem are also threatened. As just one simple example, consider the effect of the decline in bees and other pollinators on an ecosystem (32–33). Without the pollinators, plants decline, and with fewer plants, other insects, birds, and mammals also decline. We therefore need to preserve not just individual species, but entire ecosystems.

Biologists have long recognized the importance of preserving habitat. To date, progress has been mixed. On the positive side, the fraction of our planet's surface that is at least officially protected has grown to about 18% of land area and 8% of oceans (34–35). Near the end of 2022, the UN Convention on Biological Diversity adopted the Kunming-Montreal Global Biodiversity Framework, which calls for the protection of at least 30% of the planet's surface area by 2030. Like many other UN goals, this one is wildly ambitious and is very unlikely to be achieved. Nevertheless, the goal should keep us heading in the right direction (36).

On the negative side, many of the protected areas are protected on paper only (37). There is substantial human activity within many of the officially protected areas, and much of the protected area is less expensive land that is not ideal habitat. Relatively few of the protected areas are connected to one another (38–40). Large segments of tropical forest in the Amazon and Southeast Asia continue to be converted to agriculture (41).

Because money for land and water preservation is always limited, we need to carefully prioritize purchases to save as many species as possible (34, 42–43). In general, it appears that larger protected areas are better than smaller ones, although priority should also be given to small, rare habitats that occur in only a few or even just a single location on the planet. We should also try to directly connect the protected areas, or at least create corridors between them with relatively low human activity. This would facilitate migration of endangered species to new areas.

We and our descendants will also need to be continuously vigilant to maintain the protections on land and ocean areas that have already been set aside. When I was 17, my father told me to tear down a building on our lot. The building was substantial, big enough to hold a good-sized truck. Nevertheless, armed with just a few simple tools, I was able to completely demolish the building within a few days. I'm not boasting of my physical prowess. On the contrary, this experience taught me a valuable lesson: it's far easier to destroy than to build. Even if I had the knowledge and skill (which I did not and do not), it would have taken many times longer to construct the building than to tear it down. My lesson applies to areas of the planet that are reserved for other species at considerable effort and cost. These areas can be prepped for human activities, with concomitant destruction of the organisms living within them, at relatively little expense in very short time. Forests can be burned or cut down; wetlands can be drained. There will always be people who can profit from the development of the protected areas. These people will vigorously advocate to undo preservations.

- - -

For nearly all of human history, human population numbers were relatively small and changed only slowly. But over about the last two centuries, the human population has exploded (Figure 1). With our current 8 billion people, we are already placing enormous stress on land and seas. And this is despite the fact that people in developing nations consume only a small fraction of the natural resources that people in wealthier nations consume (8, 13). As developing nations advance, they will likely require more land and ocean per person.

Perhaps, through development of new technologies such as food production within buildings, our planet may be able to support more than 10 billion people and still permit other species to thrive. But we have not yet achieved such technology. Until then, I think we need to gradually reduce human populations.

Human populations in many nations of the world have already leveled off or begun to decline. Many young adults are having few or no children. Over the next 75 years (2025–2100), population decreases are projected for many nations, including China (46%), Japan (39%), Italy (35%), and Mexico (10%) (44).

Many view population reduction as a sign that a nation or even our species is in decline. I completely reject this view. If done gradually, population decreases can be achieved without reductions in standards of living. Intentional, managed human population reduction on our planet is a hopeful sign for our species, not a reason for despair. What is really important is the quality of people's lives, not the total number of people on the planet.

It's also important to note that the preservation of other species may be the most urgent of all of my suggested long term goals. Once species become extinct, they are likely lost forever. Yes, new species can evolve to replace the ones that are lost, but this takes a very long time. The celebrated biologist E. O. Wilson estimated that it would take five million years for natural evolution to replace the species that we are currently destroying (23). Attempts are also underway to resurrect extinct species like mammoths and dodo birds using sophisticated biological and genetic engineering techniques. However, no one has yet succeeded in this endeavor, and even if it can be achieved for a few marquee species, I don't see that we will have, over the next few centuries, anything close to sufficient resources to bring back all the species that are lost.

A reason for hope is that most species will recover if given sufficient space and time (45). In 1941, the population of whooping cranes, a beautiful, tall white bird native to North America, had dwindled to about 20 individuals. This compares to a population estimated at 10,000 or more prior to the arrival of Europeans. Due to intensive efforts to save this species, there are now about 700 birds in the wild in three populations, and their numbers continue to increase (46–47).

TARGETS

- Continue at least 80% of the species living today.
- Set aside at least 50% of the planet's land and ocean area exclusively for other living things (23, 48–49).
- Of this 50% of land and water, at least half (25% of the total area) should be in large contiguous reserves.

- Gradually reduce the human population living on Earth to four billion. This target would require population declines of only about 0.2% per year, considerably less than the 0.5% per year or larger declines that are currently occurring in many countries.

ACTIONS WE CAN ALL TAKE

- Contribute money, even small amounts, to preservation of additional land and water area. Many nonprofit organizations dedicated to this task already exist and can receive these donations.
- Reduce fossil fuel usage.
- Reduce meat consumption, especially red meat.
- Limit purchase of new clothing and other manufactured goods.
- Vote for political leaders who will support preservation of habitat and continuation of other species.

Chapter 4

Establish a Democratic World Government

WHAT

This goal is to replace the current United Nations (UN) with a new world government that has the following properties.

Table 5
Properties of the new world government

- All nations of the world and all people are eventually included.
- The world government has real authority—including the ability to tax, adopt laws and rules, and settle international disputes—but this authority is shared with national and local governments.
- The world government guarantees fundamental human rights for all people in perpetuity. These include the right to choose government leaders in free and fair elections. They also include other basic freedoms such as freedoms of speech, religion, the press, just

and speedy trials, and peaceable assembly (1). Totalitarian governments are eliminated.
- Representation and voting rights in the world government are based solely on population. Every person everywhere in the world has the same voting rights. Nations with the largest populations have the most votes, and nations with the smallest populations have the fewest votes.
- Uncompromising protections are woven into the fabric of the government to prevent the usurpation of human rights by power-hungry individuals. There must be no backsliding of nations (or the world) into totalitarianism.

WHY

For well over two centuries now, *many* leaders from around the globe have recognized the need for a world government to promote peace among nations and to accelerate human progress (2–6). They agree that the only way to end the warfare that has plagued humanity for all of our existence is to establish an effective world government to resolve disputes among nations. This world government would eliminate the need for substantial military spending.

Here are a few of my favorite quotes from leaders regarding world government (5):

> I believe at some future day, the nations of the earth will agree upon some sort of congress which will take cognizance of international questions of difficulty and whose

> decisions will be as binding as the decisions of our Supreme Court are upon us.
>
> Ulysses S. Grant, Civil War general and president of the US, 1877

> The future peace, security and ordered progress of the world demand a federation of free nations, and on no other basis can the problems of the modern world be solved. Such a world federation would ensure the freedom of its constituent nations, the prevention of aggression and exploitation by one nation over another, the protection of national ministries, the advancement of all backward areas and peoples, and the pooling of the world's resources for the common good of all.
>
> Mahatma Gandhi, 1942

> Mankind's desire for peace can be realized only by the creation of a world government.
>
> Albert Einstein, 1945

> Internationalism does not mean the end of individual nations. Orchestras don't mean the end of violins.
>
> Golda Meir, prime minister of Israel, circa 1976

The first major attempt at a world government was the League of Nations, established in 1920 (3–4, 6).

The League did not include all nations. The United States never joined, and the Soviet Union joined in 1934 but was expelled in 1939. The League failed to prevent World War II and is generally considered a failure. However, it did serve as a foundation for the current UN.

The UN was established in 1945 and can point to many notable successes. Its agencies, programs, and staff have been awarded the Nobel Peace Prize twelve times. The World Health Organization (WHO), a UN agency, spearheaded the global eradication of smallpox and is close to eradicating polio. As described in Chapter 1, the UN has established and is working toward many global development goals.

Although the UN is far better than no world government at all, it also has major flaws (6). The UN has no appreciable military and cannot effectively resolve international disputes and prevent war. Totalitarian governments have seats. The UN has no authority to tax and cannot pass binding legislation. There is no significant international court system to accompany the UN. The five permanent members of the UN Security Council (China, France, Russia, the United Kingdom, and the United States) have absolute veto power over any Security Council resolutions. These permanent members were set in the aftermath of World War II, about 80 years ago, and don't currently reflect world distributions of population or sizes of economies. Votes in the UN General Assembly are not based on population; for instance, the tiny nation of Andorra, with

80,000 people, has the same voting power as India, with 1.4 billion people. While the UN must be retained until a replacement is in place, it is obviously limited.

There are individuals and organizations working today toward a better world government (see, for example, 7–10). Some progress has been made. The European Union was established in 1993, and the International Criminal Court in 2002. However, overall momentum toward an effective world government is currently low. World wars provided the impetus for the establishment of world governments. The League of Nations was created after World War I, and the UN after World War II. Clearly, the horrors of world war motivated people to act. Let's hope that it does not take another ghastly world war before progress is made toward an improved world government.

Why the emphasis on democracy and human rights in the proposed world government? Because democratic governments, despite their imperfections, are the best forms of government that have yet been devised. Dictators are fond of claiming that their governments are superior to democracies, but these are just more lies from people who lie continuously. History has shown that dictators start far more wars, cause far more deaths, and are generally far worse for the welfare of people than freely elected leaders. Here are just a few relatively recent examples (11): Stalin's gulags, farm collectivization, and other atrocities killed about 20 million, mostly in Russia and Ukraine, between 1928 and 1953. Mao Zedong's Cultural Revolution and other upheavals in China killed an estimated 40 million, mostly from famine, from 1949 to 1976. Pol Pot's thugs killed 1.7 million Cambodians from 1975 to 1979, including about a third of the country's men. The string of dictators in North Korea (currently Kim Jong Un) have developed nuclear and other powerful weapons at the expense of their impoverished people. At least 0.5 million North Koreans died of famine between 1994 and 1998. Vladimir Putin's war in Ukraine, which started in 2022 and is still ongoing, has already killed or wounded over 2 million people. A large fraction of all the killings that humanity has suffered throughout history have been at the hands of dictators. Dictators are bad news. They have to go.

> Earth is a very small stage in a vast cosmic arena. Think of the rivers of blood spilled by all those generals and emperors so that, in glory and triumph, they could become the momentary masters of a fraction of a dot.
>
> Carl Sagan, 1994

HOW

Prior to the development of agriculture about 12,000 years ago, most humans were organized into bands and tribes of no more than a few hundred people (12). In many cases, bands just a few kilometers apart were mortal enemies who regularly raided and killed each other. With agriculture came chiefdoms and kingdoms with thousands of individuals. Wars were then fought between kingdoms. Gradually, kingdoms gave way to states and nations with hundreds of thousands or millions of people. Most of today's nations came into being just within the last two centuries. Now, nations fight one another. I think we need to understand this history, and understand that all boundaries between groups are artificial. All people, everywhere around the globe, are members of one single species (13). People everywhere need to work together to achieve a brighter human future.

I think that one vital step toward an effective world government is the development and adoption of a universal human language. When people from any part of the globe can easily communicate with each other, world unity will be bolstered. Although software for translating documents has improved dramatically (14), software for verbal translation is cumbersome. Verbal communication is still very important.

The inability to communicate breeds misunderstanding and distrust. It is a major component of racism. For instance, in the US today there is great controversy regarding the immigration of Spanish-speaking individuals. Native-born Americans would generally be much more welcoming to Latin immigrants if they spoke fluent, unaccented American English.

English is now the predominant international language (15–17). It is the language of science, international travel, and global finance and commerce. When scientists from around the world hold international conferences, the conferences are conducted in English.

When Dutch pilots fly over German, Hungarian, Romanian, Turkish, Iraqi, and Iranian airspace to land in Dubai, the pilots and air traffic controllers all speak English. When a Chinese businessman concludes a deal with a French saleswoman in Kuala Lumpur, they speak English. Despite the withdrawal of England from the European Union, representatives to the union usually use English to communicate with one another. Over half the people in Europe now speak English at least reasonably well. Perhaps English will become the universal world language without organized effort.

But English also has significant drawbacks. English is a chaotic amalgam of Celtic, Latin, Germanic, Scandinavian, and French languages and has borrowed thousands of words from additional languages. The spellings of English words often do not match their pronunciations, requiring precious hours of childhood (and adult) education to be devoted to memorizing spellings and pronunciations. Studies have concluded that English is more difficult to learn than languages with regular spellings (18–19), and that dyslexic children have more trouble with English than other languages (20–21). In addition, anyone around the world who has ill feelings toward England (or probably the US) will resist English. Consider French speakers in the Canadian province of Quebec who have fiercely resisted adopting English despite the fact that it is the primary language of the remainder of the country.

Personally, the universal language I favor is a phonetic version of English. Phonetic means that each letter or combination of adjacent letters has one and only one pronunciation. In a phonetic language, if a word (other than a proper name) is pronounced correctly, then that word can be correctly spelled. Children (and adults) learning the language would still need to acquire the rules for translating sounds into letters, but there would be no need to memorize spellings. This would make it easier for children to learn to read and would free substantial time in schools that could be devoted to more important subjects.

Admittedly, the record for constructed universal languages is unimpressive (16). Esperanto (1887), Ido (1907), Interlingua (1951), and others have been developed and promoted without any appreciable long term success. Implementing any universal language will certainly be challenging. Many will resist the difficulty of learning a new language. Many will oppose it because they feel it will weaken their own culture and may result in the abandonment of their own language. Demagogues will oppose a universal language to win popularity. I do not seek the loss of native languages and cultures, nor the succor of demagogues, but I still feel that the need for people from all over the world to effectively communicate far outweighs the negative aspects of a universal language. A universal language does not mean that all other languages need to be abandoned.

- - -

I submit four ideas for the gradual elimination of totalitarian states. The first would be to start the new world government now, but include only willing democratic nations. Totalitarian states would be welcome to join, but only on condition that they eliminate their dictators and adopt democracy. The relentless message to *people* in totalitarian states would be

> "Come, join us in peace, freedom, and prosperity."

and

> "Be part of the bright human future, not the dark human past."

Second, to emphasize the prosperity anticipated in the above slogan, trade between the free states and the totalitarian ones should gradually be reduced. Scientific and technological knowledge transfer should also be restricted. This would unfortunately cause economic hardship among the people living in totalitarian states, but would also put pressure on the dictators. The message to people living under such sanctions should be:

> "Join us, and the sanctions will be lifted."

Third, the new world government would also be a mutual compact for protection and defense. An attack on one would be an attack on all. Dictators who start wars would immediately be charged with the crime of murder. About 2,500 years ago the Chinese philosopher Mozi said (22):

> To kill one man is to be guilty of a capital crime, to kill ten men is to increase the guilt ten-fold, to kill a hundred men is to increase it a hundred-fold. This the rulers of the earth all recognize, and yet when it comes to the greatest crime—waging war on another state—they praise it!

Fourth, free nations and peoples need to begin an aggressive truth offensive. Truth will rarely have immediate effects, but over the long run, it is an extremely powerful force. Dictators fear truth. That's why they always control the media. Free peoples need to defeat these efforts. International communication has never been easier than it is now. The information barriers erected by dictators always leak. The free nations of the world need to exploit and expand this leakage. A significant fraction of the money now spent by free nations on weapons needs to be shifted to programs to deliver the truth to people under the thumbs of dictators. A universal language will also, of course, help considerably in propagating the truth.

None of these approaches by itself is likely to be sufficient, but if all are applied continuously for decades, then I think ultimate success will be achieved. After all, people want to be free.

- - -

Some oppose a true world government because they fear it would lead to a global dictator and the loss of freedom everywhere. Multiple, layered protections need to be woven into the charter of the world government to prevent individuals or small groups from obtaining too much power. Irrevocable term limits for elected officials and limits on campaign financing must be included. Ultimately, however, maintaining freedom is the responsibility and duty of ordinary citizens. I doubt there will ever be any shortage of power-hungry individuals who will be happy to take away people's freedom and become dictators. *Every* new generation must therefore fight to maintain their freedom.

> History does not long entrust the care of freedom to the weak or the timid.
>
> Dwight D. Eisenhower, World War II general and US president, 1953

TARGETS

- Approximately 55% of the world's population currently live in free (or mostly free) nations (23–24). Increase this percentage to 100%.
- By 2040, shift at least 10% of current spending by free nations on militaries to a broad truth offensive.
- Establish a new world government of democratic nations by the year 2075 at the latest. From there, strengthen and expand it to include all of the properties listed in Table 5.
- Today, roughly 20% of the world's population has at least a rudimentary ability to communicate in English. Increase the fraction of the world's people who fluently speak a universal language by at least 3% per decade.

ACTIONS WE CAN ALL TAKE

- Strive relentlessly to achieve and maintain freedom.
- Learn a universal human language.
- Vote for political leaders who will support a world government.

This goal may be the one for which *individual* action is most important.

Chapter 5

Manage the Human Gene Pool

WHAT

This goal is to minimize in the human gene pool the number of deleterious DNA sequence variants that increase the risk of disease or diminish our abilities. I define the human gene pool as the sum of all DNA molecules in all people living at any given time.

This goal is appreciated today by very few, yet in my opinion is one of the most important. If we don't address the problem that underlies this goal, it's likely that human societies will gradually crumble.

Because appreciation of this goal requires some knowledge of genetics, this chapter may be more difficult for readers to follow than the others. I refer those interested in learning more about this topic to another book that I have written (*Genomes and the Gene Pool*) (1). What follows is just a short summary.

WHY

The DNA molecules that comprise our genomes have a large influence over our health and abilities. While environment is also very important, the major role of DNA is a scientific fact supported by mountains of evidence.

DNA molecules are passed down from parents to children, generation after generation. While our bodies last at most about 100 years, DNA molecules continue for thousands and millions of years with only slow, gradual change.

The process of mutation slightly changes the DNA molecules that are present in a child compared to the DNA in the parents. Many people think of mutation as a rare, freakish event that occasionally produces a person who looks or acts very differently from the rest of us. In reality, however, mutation is a natural and ubiquitous biological process. *Every* child that is born has roughly 1,000 mutations.

The differences in people's genomes that arise through mutation are called DNA sequence variants, or variants for short. In terms of their effect upon our health and abilities, variants can be deleterious, neutral, or advantageous. The great majority of new variants from mutation are neutral—that is, they have little, if any, effect on our health and abilities. However, of the new variants that are not neutral, many more are deleterious than advantageous. These deleterious variants increase the risk for, or directly cause, thousands of different disorders affecting all of our bodies' tissues and organs. They can, for example, cause heart, kidney, bone, or skin disease. They can also cause dysfunction of our eyes or muscles, limiting our ability to see or move. And they can affect our brains, leading to autism, epilepsy, or intellectual disability. Deleterious variants range widely in impact. Some are incompatible with life, some cause major disease or disability, and some only slightly diminish our health and abilities.

Since many more mutations are deleterious than advantageous, what prevents deleterious variants from accumulating in genomes over multiple generations? How can humans or any other species even survive, much less thrive, over long periods of time? The solution to this mystery, called natural selection, was discovered independently in the mid-19th century by two British biologists, Charles Darwin and the lesser-known Alfred Wallace (2–3).

Prior to the last few centuries, with few exceptions, life for our ancestors was extraordinarily difficult and harsh. Dangers such as famine, infections, violence, storms, and predators were ever present, and most lives were short. In general, those who through mutation had poorer health or diminished abilities reproduced at lower rates than those who were healthier and more talented. Darwin called this the continuous "struggle for existence." Others called it "survival of the fittest." Through natural selection, deleterious variants were kept in check.

Largely through advances in science and technology (Chapter 6), humanity has substantially relaxed natural selection. Over about the last 200 years, the average human lifespan has approximately doubled (4–6). While reductions in both maternal and adult mortality have contributed to the doubling, by far the largest factor has been massive drops in infant and child mortality. In most preindustrial human societies, at most times, 30%–50% of all newborns died before reaching reproductive age. Today, in developed nations, the rate is well under 1%.

This huge drop in infant and child mortality is a marvelous advance for humanity, but it has also resulted in a *profound* change in human evolution. Prior to this drop in mortality, for millions of years, many children with sequence variants that impaired their health and abilities died before they could reproduce. The deleterious variants were thereby removed from the gene pool. Today, nearly all of these children survive, and many pass the variants on to their own children. Deleterious variants are therefore accumulating in the gene pool. Average innate health, vigor, and intelligence are all slowly declining. This is the gene pool problem. If we take no action, then at some point, the effect upon human societies will become devastating.

How long before the effects of the degradation of the gene pool begin to be felt? No one knows. The problem is too complex and has been studied far too little. It is quite possible that the gene pool problem is already affecting our societies. I do know, however, that like climate change and many other problems, the longer we wait to address the gene pool problem, the more difficult it will be to solve. Mutation is relentless; it never rests. Complacency in this matter is reckless. If we want human societies to thrive into the future, we *must* solve the gene pool problem.

Please understand that I am not in any way attempting to demean those who are unhealthy or have weaker minds. No one has any control over their own genome. Your genome is just something with which you are born.

HOW

One important step in addressing the gene pool problem is to break down the taboos that currently surround this topic. Many people today consider any mention of the gene pool to be immoral or even malevolent. I think these taboos largely arose as backlash against the Eugenics Movement of the first half of the 20th century. During the Eugenics Movement, roughly 600,000 people worldwide, mostly in Nazi Germany, the United States, and Sweden, were involuntarily sterilized (7–9). In Nazi Germany, the movement was taken to a gruesome extreme. The Nazis murdered an estimated 250,000 people whom they considered genetically inferior (9).

The Eugenics Movement was tragically flawed because it was racist, degraded the weak, was based on limited and inaccurate knowledge of genetics, and employed involuntary sterilization as the primary means of accomplishing its goals. It was a dark episode in human history that hopefully will never be repeated. But it would be a huge mistake to let the Eugenics

Movement paralyze us forever from any consideration of the gene pool. We need many of our best minds to study and analyze the gene pool. We need to develop effective methods to monitor changes to the gene pool over time. And we need to devise humane, practical, and effective solutions to the gene pool problem.

Since we have greatly weakened natural selection, we can no longer rely on nature to keep mutations in check. Instead, we must actively manage the gene pool. Those who are fortunate to be healthier and more talented must reproduce at higher rates than those who are less fortunate. Voluntary approaches can't hurt and might even be sufficient, but it may also be necessary for governments to introduce incentives and/or disincentives to achieve the needed results. Further discussion of possible solutions to the gene pool problem can be found in *Genomes and the Gene Pool* (1).

There are risks inherent in any gene pool management program. Abridgment of reproductive freedom is one of these risks. Racism is another. I encourage people to absolutely reject any proposed program that does not at a minimum abide by the restrictions shown in Table 6. We must act, but we need to proceed cautiously and only after extensive study and discussion.

Table 6
Restrictions for programs that address the gene pool problem

- Reproductive freedom must not be abridged. Involuntary sterilization or involuntary abortion, for example, must never be employed.

- Any reproductive incentives (or disincentives) must scrupulously avoid being based in any way on race, religion, political views, or any other factor that would encourage discrimination.
- Since optimal criteria to determine eligibility for incentives are unknown, different nations and regions should adopt their own independent plans, rather than implementing universal plans that cover hundreds of millions or billions of people.
- Plans must be developed only in democracies and must be implemented only after extensive discussion and debate.

TARGETS

- Increase the annual number of research publications on the gene pool by a factor of at least 1,000.
- Devise effective methods to monitor the gene pool.
- Stop the current accumulation of deleterious variants in the gene pool.

ACTIONS WE CAN ALL TAKE

- Speak out on the gene pool topic; do not be muzzled by "political correctness."
- Have your own genome sequenced.
- Take responsibility for your own reproduction.
- Work relentlessly to stamp out racism.

Chapter 6

Grow Knowledge

WHAT

This goal is to increase as rapidly as possible the collective stock of human knowledge. Under the umbrella of knowledge, I include data, technology, facts, and understanding.

As I define them, there is considerable overlap among the terms "data," "technology," "facts," and "understanding," but they are still distinct. Data are collections of such items as measurements, observations, and responses to questionnaires. Technology is the know-how that is used to produce all products and services, everything from food to medicine to spacecraft. Facts are truths about our natural world and about human activities. Understanding is a deeper appreciation of a subject that goes beyond mere facts. Charles Darwin amalgamated many facts to achieve his breakthrough understanding of the evolution of species (1).

WHY

Humanity has advanced tremendously, especially over about the last 200 years. Agricultural yields for many important crops have more than quadrupled (2). Average life expectancy has doubled (3). Fabulous new technologies have been developed. Average per capita income, which was pretty much unchanged for millennia, suddenly zoomed (4–5). No other 200-year period in the 200,000-year history of modern humans has seen anything even remotely close to this level of progress.

This progress has not come because of great changes in government, law, religion, or social science. All of these areas have seen some improvement, but those advances have been dwarfed by advances in natural science. Science and the technology derived from science have achieved astounding progress over the last 200 years (4–7). Compared to two centuries ago, humanity now has a vastly better understanding of our natural world.

Investment in science has turned out to be among the most lucrative financial investments of all time. Studies have reported returns on investment in basic research of 20%–60% or more per year (8). The Human Genome Project, the organized effort to sequence the first human genome, was completed in 2001 at a cost of approximately 3 billion US dollars. Yet the return on this investment through 2012 was reported as nearly 1 trillion dollars, along with the creation of 4.3 million job years (9). Nearly all large private companies invest heavily in applied research and development. They do this not out of altruism, but because business leaders know that such investment will improve the company's competitive and financial positions.

Humanity has gained so much scientific knowledge over such a short period of time that sometimes people think there is not much more to learn. Even accomplished scientists sometimes fall into this trap. In reality, however, despite our progress, our current cumulative knowledge is like a tiny, brave flame in a vast ocean of darkness. Humanity still has abundant problems and needs. We desperately need to continue to learn.

HOW

Growth in knowledge is the one long term goal in this book toward which we are currently making good progress. Scientific research can always be expanded and improved, but more than anything else, we just need to continue our current trajectory. It is absolutely essential that nations continue to spend on research. And the spending must be devoted not just to applied research that is targeted toward solving specific problems, but also to basic research intended to increase overall knowledge and not to address any specific problem.

We need to continue to spread cutting-edge research to all parts of the globe. Just a few decades ago, nearly all leading research was carried out by scientists with European ancestry. Talented scientists from other parts of the world often made significant contributions, but these usually involved traveling to and working within research groups led by Europeans. However, especially over about the last 20 years, I have seen this situation change. Many more significant advances are now being made by scientists outside of Europe, North America, and Australia (10). Even relatively poor, developing nations should devote nontrivial resources to research. In selected research areas, they will be able to compete with large nations, and the development of scientific talent will pay off handsomely.

The results of scientific research need to be made readily available to scientists and the general public throughout the world. Scientists always have and always will stand on the shoulders of their predecessors. When I started in scientific research in about 1975, the only ways to learn about the latest scientific developments were to read *printed* journal articles, attend seminars delivered in person by distinguished scientists, or physically travel to conferences. The first two avenues were available only at major research universities. The third avenue required substantial funds, which many scientists from developing countries and most students everywhere did not (and still do not) have. Today, however, many journal articles, seminars, and even sometimes conferences are available for free or at modest cost to anyone with good internet access (11). We need to continue to expand this process of "open access," so that any budding young scientist in, say, Uganda or Bolivia or Turkmenistan can easily view the latest, greatest scientific advances.

We need to entrench peer review in scientific research in all nations (12–13). Peer review means that scientists themselves determine which research proposals are funded and which manuscripts are accepted for publication in the best journals. I have personally participated in many peer reviews. The process is not perfect. Human prejudices always intrude. But overall, the peer review process is the best approach that anyone has found to advance science. The alternative is to let bureaucrats, or even worse, ideologues, decide who receives the research funds. This inevitably leads to funding of the scientists who are most subservient to the decision-makers. The end result is shoddy research.

Scientists need to be shielded from the interference of political and other leaders. Sometimes scientists uncover new knowledge that conflicts with the agendas of leaders. Think, for instance, of Galileo, who was forced by the Catholic Church to recant that the earth orbits the sun, or climate scientists who have been muzzled by politicians who find it advantageous to support the fossil fuel industry. Scientists need the freedom to pursue and voice the truth, regardless of where it may lead.

Scientific education at all levels needs to be expanded. Vast numbers of people still hold beliefs that are contradictory to overwhelming scientific evidence (14–15). Examples include denial of human-induced climate change and great exaggeration of the risks of vaccines. Such beliefs are usually harmful and can be disastrous. No one has found a completely effective method of weaning people from such beliefs. However, I'm convinced that expanding understanding of science and the scientific method will help. Expansion of scientific education will have the additional benefit of priming the pipeline with future scientists and engineers.

TARGETS

- Increase the fraction of world nations that are devoting at least 3% of their GDP (gross domestic product) to research and development from the current approximately 5% of nations (5, 16) to at least 90% of nations.
- Make all scientific research publications freely available immediately to anyone, anywhere, who has internet access (or the future equivalent).
- Make peer review, as opposed to cronyism, the primary basis for allocating public research funds in all countries.
- Double the number of hours of required STEM (science, technology, engineering, and mathematics) education for students at the grade school and high school levels, and for nonscience majors at colleges and universities. Emphasize teaching of scientific methods.

ACTIONS WE CAN ALL TAKE

- Try to understand the scientific process.
- Learn about a few areas of current scientific progress that seem interesting. A basic level of understanding is definitely possible.
- Contribute funds to scientific research.
- Support STEM education.
- Oppose attempts to muzzle scientists or to diminish scientific freedom.

Chapter 7

Temper Greed

WHAT

This goal is to achieve a modest reduction in self-interest. The goal is not some totally unrealistic aim to convert people into saints, but rather to make reasonable and valuable progress.

I considered several alternative titles for this chapter, including "Decrease Inequality," "Improve Morality," and "Reduce Injustice." There is considerable overlap among all of these, and all are certainly desirable, but I think reduction of greed best summarizes my intention.

WHY

One of the greatest checks on human progress has always been and continues to be our strong self-interest. It's exceptionally difficult to get people to give up even a few of their possessions or comforts for the common good. Regardless of how much we possess, we always want more. This is our nature.

There are countless examples of how greed holds us back, but to list just a few, consider that investors prefer to develop every single hectare of land, rather than reserve even a small portion for other species. People drive large, fuel-inefficient vehicles, even though they contribute excessively to climate change. Business leaders award themselves enormous salaries while curtailing the wages and benefits of their employees.

Over the ages, many scholars have considered the question: what makes a human society successful? (see 1–6 for a very brief selection). I'm convinced that there is no simple answer to this important question. Success of a human society depends upon a complex integration of many different factors. One of these factors, one which I think has been generally underestimated, is the morality of the society.

> I grew convinced that truth, sincerity, and integrity, in dealings between man and man, were of the utmost importance to the felicity of life.
>
> Benjamin Franklin, *Autobiography*, 1784

Many economists have concluded that corruption in business substantially reduces economic growth and national wealth (see, for example, 7–10). The flow of business is lubricated by a basic level of trust. If business people routinely try to cheat each other, then economies are crippled. It is therefore no surprise that countries with the highest levels of corruption tend to be the poorest.

Morality is also an essential component of democracy. The founding fathers of the United States recognized this. Consider the following quotes:

> Is there no virtue among us? If there be not, we are in a wretched situation. No theoretical checks–no form of government can render us secure.
>
> James Madison, 1788

> Virtue or morality is a necessary spring of popular government.
>
> George Washington, 1796

Economic inequality encourages distrust and resentment and erodes democracy (11–12). There is growing evidence that well-being and happiness are promoted by reducing inequality (13–15). In particular, those with the lowest income and net worth may become substantially happier through even modest reductions in inequality.

HOW

I think we can nearly all agree that tempering greed is an attractive goal, but achieving it is extraordinarily difficult. Huge numbers of people have tried over millennia without much success. I don't pretend to have anything even close to a full solution, but I offer a few ideas that may help.

To my knowledge, all major world religions influence human behavior through the promise of reward or punishment after death (16–17). The monotheistic religions, Christianity and Islam, have heaven and hell. If you live a moral life, then after death you go to a heavenly paradise. If you are bad, then you suffer eternal torment in hell. Similarly, in Hinduism, if you live a good life, you will reincarnate into a higher caste, whereas if you are bad, you might reincarnate into the lowest in society.

I suggest that we continue this proven approach, but move the reward and punishment from after death to retirement age. Using set criteria, governments would record and store the performance of each citizen. Then, those who have been "good" in life would receive greater old-age pensions and healthcare benefits, and those who have been "bad" would receive lower payments and benefits. The rapid development of information technology makes tracking life performance practical.

Many different criteria could be used to set the retirement payments in this "heaven on earth" concept. As a starting point, I propose three: employment, voting, and criminal records. Those who have been gainfully employed their entire working lives, have voted in the most elections (regardless of how they voted), and have not been convicted of any crimes would receive the highest payments and benefits.

I don't know if these three criteria are the best to encourage moral behavior. There are certainly many other possibilities. Some criteria would clearly be easier to track and evaluate than others. I think that the list of criteria could begin relatively short, and then be gradually expanded over decades and centuries. I have provided an expanded list of my personal preferences for criteria in Appendix 1.

As additional features of the "heaven on earth" plan, I suggest that individual rankings be required to approximate a normal distribution. Most people would receive retirement benefits that are close to the average. Few would receive the highest or lowest payments. The difference between the highest and lowest payments would be sufficient to influence people's behavior, yet not be totally lacking in compassion for those who have struggled. In addition, those who improve their behavior later in life should be able to compensate for mistakes made when young. This will reduce the number of people who abandon efforts to improve. Finally, those relatively few whose behavior worsens after retirement would see their benefits drop. Monitoring of performance would continue until death.

- - -

I also propose that the government records on individual performance be made freely and publicly available through the internet at retirement age or death, whichever comes first. In this way, family, friends, neighbors, and all others will be able to view the records of anyone who has reached retirement age or died. The records would be retained forever and would contain basic facts about the person, such as birth and death dates, addresses, and records on voting, employment, and crime. Hopefully, publication of such records would serve as additional motivation for people to improve their actions. People would understand that anyone, stretching far out into the future, will be able to learn what type of life they led.

Further, I suggest that individuals be allowed (but not required) to prepare their own life summaries of, say, 1,500 words or less, which would accompany the government records. These summaries could be prepared only by the individuals themselves and could contain whatever information they want. These summaries would also be retained forever and be freely available at retirement age to anyone who wishes to look. An image or video of the person could perhaps be added. As an example of what I have in mind, my own summary is included in this book as Appendix 2.

The primary purpose of the public records would be to motivate people to live better lives, but another valuable purpose would be to help us learn about our ancestors. I think most, maybe nearly all, people are curious about their ancestors and desire to learn more about them. In my case, my paternal grandfather died 20 years before I was born. Although he contributed 25% of my DNA, I didn't know him at all. I did know my other three grandparents, but only as a child. They died before I could become acquainted with them as an adult. I would dearly like to learn more about my grandparents and more distant ancestors. The basic life records and the life summaries would help.

- - -

Another idea to temper greed is to require every organization with, say, more than 100 employees, both for-profit companies and nonprofit organizations, to report annually the ratio of total compensation for the 5% of full-time employees who are paid the most to the total compensation for the 5% percent of full-time employees who are paid the least. For instance, if the total compensation for the most highly paid 5% was $3,000,000, and the total compensation for the lowest paid 5% was $300,000, then the ratio would be 10. Since essentially all organizations routinely track compensation, this ratio would be easy to calculate. Available records might even allow governments to verify the numbers.

I further propose requiring that this inequality ratio be prominently displayed on *all* organization websites and in *all* organization advertising. This would allow consumers, employees, and donors to evaluate organizations on the basis of fairness. Of course, other factors would still dictate many decisions, but I think many would take the ratio into account. Success of the organization would then depend in part on the ratio.

- - -

Government taxation can also, of course, play a role in reducing inequality. There appear to be three main methods of reducing economic inequality through taxation: progressive taxation of income, taxation of wealth, and taxation of inheritance (18–20). Although to a degree I favor all three approaches, I particularly endorse taxation of inheritance. I see a need to reward effort and success to maintain a healthy economy, but I don't see a need or justice in allowing children to inherit huge amounts of wealth that they didn't earn (21). I favor a strict cap on the amount of wealth that can be inherited.

My guess is that an effective world government (Chapter 4) will facilitate reductions in inequality among nations (as opposed to within a nation). Collection and retention of economic data for all nations will be necessary to measure progress toward this goal (Chapter 8).

We should not accept the idea that humanity is doomed to forever suffer great inequality. Reductions in inequality are definitely possible. In the years after World War II in North America and Europe, inequality in wealth and income *within* nations was substantially lower compared to previous centuries and compared to the present (20, 22). Also, inequality *among* nations appears to have been substantially lower before the Industrial Revolution than it is today (23).

TARGETS

- Cut average organizational *compensation* inequality, as defined by my wage-inequality ratio (top 5% divided by bottom 5%; see above) to at least half of its current level. Determine whether this can be voluntarily achieved by people "voting" with their expenditures and their employment.
- Cut *wealth* inequality *within* nations, averaged over all nations, to at least half of its current level. Possibilities for measuring inequality include Gini indices (24) and the ratio of the total net worth of the

wealthiest 5% of a country's population to the total net worth of the poorest 5%.

- Cut *wealth* inequality *among* nations to at least half of its current level. The goal here is to reduce the gap between the wealthiest and poorest nations. A possible metric is the ratio of the average net worth per capita of the richest 5% of nations to the poorest 5% of nations.
- Test the "heaven on earth" and "life summary" approaches in at least a few populous nations for at least a century. Adopt widely if successful. Assessment of success may include trends such as reduced inequality, higher voting rates, greater employment, and less crime.

These targets may seem overly ambitious, but we need to remember that inequality is currently very high. Cutting inequality in half over 300 years will still leave our world a very inequitable place.

ACTIONS WE CAN ALL TAKE

- Consider carefully how your own decisions and actions will be evaluated by succeeding generations.
- Support companies and organizations that are more equitable through your purchases and employment.
- Vote for political leaders who will vigorously strive to reduce inequality.

Chapter 8

Preserve Knowledge

WHAT

The goal described in Chapter 6 was to *grow* knowledge. The goal in this chapter is to *preserve* knowledge. Critical knowledge must be retained in perpetuity to maintain and advance human development. This goal has four components, as shown in Table 7.

Table 7
Components of the preserve knowledge goal

- Prioritize the knowledge that needs to be retained in each field of human endeavor.
- Save the knowledge identified through the prioritization step as far as resources permit.
- To the greatest possible extent, make the knowledge readily accessible to all people throughout the world.
- Secure knowledge repositories against unintentional or intentional loss.

Using the definition presented at the beginning of Chapter 6, knowledge includes data, technology, facts, and understanding. Preservation of all of these is important. However, the volume of data is orders of magnitude greater than the volume of technology, facts, and understanding combined. For instance, many of the basic facts and understanding about genetics can be found within a few authoritative textbooks, but the genome of a *single* individual, if printed, would fill roughly 4,000 such textbooks, and eight billion human genomes currently exist. Because of the enormity of data, it will not be possible to preserve all data; hard decisions about retention will need to be made.

WHY

The development of written languages and the creation of libraries as repositories of knowledge were critical milestones in human development. But unfortunately, past human societies did not sufficiently value libraries, and so huge troves of knowledge were lost (1). As perhaps the best-known example, the Great Library of Alexandria in Egypt housed a large collection of writings from classical Greece and other civilizations, yet these writings were almost entirely lost during the early years of the Roman Empire (2–3).

Absence of knowledge impoverishes our lives. For example, many would like to know how modern humans and Neanderthals interacted when they came together. We know that they at least occasionally reproduced together and that Neanderthals became extinct relatively soon after contact with modern humans, but beyond that we know nothing. From the writings of classical Greeks, we know that they developed a rich culture of music, yet we have no idea how it sounded.

Loss of knowledge also impoverishes our lives. Imagine what our lives would be like without the Bhagavad Gita, the philosophies of Mozi, the music of Beethoven, or the novels of Dickens. The primary purposes of studying history are to avoid repeating past mistakes and to capitalize on past successes. Without knowledge of history, our lives would be burdened with unnecessary struggle and suffering.

The loss of past knowledge forces people to reacquire the knowledge at great cost. It greatly slows progress. If scientists, for instance, had to rediscover the laws of nature over and over, the pace of scientific advance would become glacially slow. Many basic concepts in religion, law, business, and government were developed in the past, often the distant past. Absence of these basic concepts would set our societies far back.

We need to retain data so that we can accurately determine trends. Is inequality growing or shrinking worldwide? Is a particular species increasing or decreasing in number? We can only answer such questions if the relevant data are retained. Monitoring all of the targets proposed in this book requires that data be collected and saved over centuries so that our descendants can assess progress toward goals.

In addition to the loss of older knowledge, much critical knowledge from today is not being retained (4–7). Since I'm a geneticist, I'll give an example from my field. Geneticists now have the capability to sequence people's genomes at modest costs, but our ability to interpret the sequences is at an early stage. To improve interpretation, both DNA sequences and health records from large numbers of people are needed. This information is useful now, and will be useful thousands of years into the future. Yet today, this information is usually not retained. To me, this is unacceptable waste. We have more than sufficient resources to save these data, but we lack adequate processes and will.

If a major catastrophe such as a massive pandemic or a supervolcano strikes, and societies break down, then to recover, people will desperately need the knowledge from previous generations. This has been a frequent theme of fiction (8–10).

HOW

Fortunately, the need for knowledge retention is widely recognized (see, for example, 11–14). Organizations and institutions have been established that are dedicated to knowledge retention (for example 15–19). Such efforts are a good start, but more is needed. We need to consider the problem from a higher altitude and longer term perspective.

I think the world and national governments need to periodically convene expert panels in all fields to make recommendations concerning retention priorities. Most people would probably agree that virtually all technology, facts, and understanding should be retained indefinitely, but data are different. Although the costs of data storage have dropped dramatically in recent decades (see below), data are also being generated today at unprecedented rates. Some data will likely be retained indefinitely, some only for a relatively small number of years, and others not at all.

I don't pretend to know which knowledge is the most critical to retain, especially in areas outside my expertise, but the following very brief list may contain some useful thoughts.

Table 8
Some knowledge to be retained indefinitely

- Basic factual records for every individual and life summaries for those who write them, as described in Chapter 7
- Widely read books
- Research journal contents
- Census data
- Financial and business information, including information on inequality
- Health records and DNA sequences of every person (for healthcare purposes)
- How-to manuals for essential processes like water treatment, delivering babies, growing rice, and producing cement and steel
- Treasured music and artwork

As has happened many times in the past, knowledge can be lost due to neglect, catastrophic events such as wars, or deliberate destruction. Book burning has happened in the past and will likely occur again. When knowledge exists only in digital form, that knowledge can be destroyed by the mere click of a button.

To prevent the loss of knowledge, I think we need to place our most critical knowledge in safe locations, probably deep underground or, eventually, off planet. Physical objects that can easily be destroyed, such as paintings, fossils, and archaeological artifacts, need to be digitized (20). In case of the collapse of civilization, it would also be wise to store critical knowledge on media that endures for long periods and that doesn't require electricity or sophisticated technology to maintain and access (21–24). Although paper doesn't last forever, printing is well established and relatively inexpensive. Printed copies of the most critical knowledge should continue to be made. The more copies in more places around the globe, the better.

Money for knowledge retention will always be limited, but centralized knowledge repositories need not be funded entirely by taxpayers (at least not directly). Many data have considerable financial value. Medical data, for instance, that are used to develop drugs, software, and devices are worth billions of dollars (25–26). It should be possible to at least partially fund the repositories through fees paid by private companies for access to deidentified data.

Great progress has been made in recent decades in lowering the cost of data storage. Since 1960, storage costs have dropped from about $30 billion per terabyte to about $10 per terabyte today, a three-billion-fold reduction (27). Over about the last 15 years, costs have continued to drop, but more slowly. Significant research resources need to be devoted to further reductions in storage costs. Every twofold drop in cost doubles the amount of data that can be retained.

Large numbers of librarians who are expert at curating and managing huge amounts of data need to be trained and deployed. Perhaps a relatively new profession, data archivist, needs to be further developed.

Wherever possible, we must make the knowledge that is retained available to all. Certainly, copyrights and patents as well as personal privacy need to be respected, but the restrictions must have bounds. In my opinion, protections for the creators and owners of knowledge have grown excessive in recent decades (28–29). We need to trim these restrictions.

TARGETS

- Convene groups of experts from around the world in all fields every 20 years to recommend priorities for knowledge and object retention.
- Increase spending in each nation for public knowledge repositories to a level of at least 0.10% of GDP. For reference, the US federal government currently spends roughly 0.01% of GDP on libraries and museums.
- Establish national libraries in all nations. Forge tight connections among the national libraries through the UN or other world government.
- Establish at least six highly secure and continuously curated repositories for knowledge and objects at six separate locations around the world. Most knowledge should be shared among each of the

repositories. The repositories must retain critical knowledge in a format that does not require electricity to maintain and access.

- Decrease the cost of data storage by a factor of at least 10.
- Train and employ at least 250,000 data archivists worldwide.
- Make useful and essential knowledge freely available to all people everywhere.

ACTIONS WE CAN ALL TAKE

- Vigorously oppose any and all attempts to destroy knowledge and objects of cultural importance. Book burning has been, is now, and forever will be foolish and wrong.
- Maintain your own printed or digital copies of books/documents that you particularly treasure.
- Support increases in government spending for knowledge and object retention.

Chapter 9

Set and Accomplish Long Term Goals

PONDERING POTENTIAL GOALS IS USEFUL, but vastly more beneficial is to actually set and then accomplish them. I think that *governments* need to officially establish long term goals, prepare plans, monitor progress, and facilitate achievement. I further assert that only *world* governments can effectively set and manage long term goals for all humanity. Ideally, the world government would be similar to the one I described in Chapter 4, but until such a government is created, the current UN will have to serve.

Legally binding processes for setting and monitoring long term goals need to be instituted. One approach would be for the charter of the world government to be amended or drafted so as to require the government to set between say, 5 and 15 long term goals every 50 years. This would work similarly to the provision in the US Constitution that requires the government to complete a census of the population every decade. To ensure that there is broad public support for the long term goals, the goals would need to be approved by a majority, perhaps a supermajority, of elected representatives of the people. There should be a minimum number of goals so that the process is not irrelevant, and a maximum number so that efforts toward accomplishment are not excessively diluted.

Because human societies are continually changing, it is difficult to imagine that a goal set, say, in the year 2050 would be entirely appropriate and relevant by the year 2350. A 50-year reset of the long term goals would allow goals that have already been met or are no longer relevant to be dropped. Other goals would need to be amended to cope with changing circumstances. Targets will often need to be adjusted. New goals can be added. Even if a goal lasts only 50 years, as long as it is worthwhile, then progress toward its achievement will not be wasted.

Once ratified, a goal and its targets could not be changed over the next 50 years. "Long term" has no meaning if the goals can easily be tampered. Of course, leaders who do not like a particular goal will not pursue it, or may even try to undo progress, but leaders in other nations and many individuals everywhere will continue to work toward the realization of worthy goals. If unworthy goals are set, then my guess is that few people will work toward achieving them, and they won't be realized.

In addition to setting goals, the world government would also be required to prepare relatively detailed plans and milestones for accomplishing each of the goals. Because plans depend upon available technology and current circumstances, the plans (but not the goals and targets themselves) will need to be updated more frequently, perhaps every 10 years.

It is also essential to monitor progress toward the goals, so that our descendants know if they are on track toward achievement and know when the goals have been reached. The world government must therefore establish a department that would be required to periodically produce a progress report on each long term goal, perhaps every 10 years. Unlike setting goals, I think that monitoring the goals should also be the independent responsibility of multiple nongovernmental organizations. This will serve as a check on inaccurate and politically slanted government reports.

To encourage the world governments to facilitate accomplishment of the goals, reports from nonpartisan government bodies should be generated to summarize the impact of *all* new world government legislation on the completion of each of the long term goals. Individual nations should also require that such reports be generated for their own national legislation. The US Congressional Budget Office is an example of the type of government body I have in mind.

- - -

Drafting long term goals, preparing plans, and monitoring progress should require *relatively* little money and time. Far greater resources, however, will be necessary to accomplish the goals. I doubt worthwhile long term goals will be reached without the combined actions of governments, nongovernmental organizations, and the public.

Governments at all levels, from the world government to local governments, can make significant direct contributions toward goal achievement. Although taxpayer money is always limited, governments often have greater resources than nongovernmental organizations.

Much of the labor needed to accomplish the long term goals would likely be provided by people employed by nongovernmental, nonprofit organizations. The numbers of such people should be increased substantially. Endowment funds of the nonprofit organizations are particularly important in this regard.

Endowment funds of nonprofits today total over $1 trillion worldwide (1–2). This is about 0.2% of total world net worth. A large fraction of the endowment total is held by universities. Roughly 400,000 people worldwide are currently employed through these endowments. This is only a tiny fraction of the world's population (0.005%), but these 400,000 people are working *full time* on the types of goals outlined in this book. Governments should encourage the establishment and growth of the endowments by making donations to the organizations tax deductible, by limiting the amounts of wealth that can be inherited so that wealthy individuals will donate to nonprofits through their wills, by keeping taxes low on the endowments' investment earnings, and by limiting the requirements for annual spending from the endowments. In return, the endowments will help to stimulate economies by providing capital. Governments can set investment rules for the endowments so that they maximally benefit economies.

For-profit companies will probably in general have less ability to assist in progressing toward the goals, but they can still contribute, particularly at times when they are especially profitable. Company employees can be encouraged to donate through matching gifts. Employees can also be given a small amount of paid time off of work *if* they devote that time toward goal accomplishment.

The efforts of individuals will also be vital to goal achievement. To start people on this path, I suggest that the goals be taught to all students in grade and high schools. I also suggest that young adults of about 18 years of age be required to devote a small portion of their lives, perhaps a year or two, in service to their nations. During their service, the young people would receive food, clothing, shelter, additional education, and healthcare, but little pay. This would hold down the costs of such programs. The labor of the youth would benefit everyone, but the young people themselves would also benefit through the encouragement of healthy lifestyles, improvement in personal discipline, and educational advancement. Perhaps most importantly, such programs would instill in the young people the concept that we all have a duty to give back to the societies that provide us with so much.

Many individuals will want to contribute to the goals but won't know where to start. Nonprofit organizations should therefore devise creative approaches to allow people to *easily* contribute in meaningful ways. Anyone who wants to help should be able to find lists of activities on organization websites. Substantial resources should be devoted to organizing these activities. Ideally, such activities should include work that can be easily performed by a person at home, as well as in groups outside of homes. Even 30 minutes per week, when multiplied by millions and billions over centuries, will have great impact.

Older people who are retired from their primary careers can make particularly important contributions. In the past, people usually either died while working or worked up until retirement age and then ceased working entirely. But today, many retirees have sufficient health and energy to contribute at least part time to long term goals.

People living today have a much greater opportunity to contribute to worthwhile goals than those who lived in the past. Consider that 500 years ago the great majority of people throughout the world were farmers. They needed to work long hours just to produce enough food to feed themselves and their families. Today, however, in developed countries, less than 2% of people are farmers. With modern technology, a small fraction of the population can provide us with the necessities of life. This frees huge amounts of time and energy that could be devoted to worthwhile long term goals.

TARGETS

- Modify the UN Charter to require the setting, planning, and monitoring of long term goals.
- Include a requirement for long term goals in the founding organizational document of the new world government.
- Increase the fraction of the world's population working *full time* toward accomplishing worthwhile long term goals from the current roughly 0.005% to at least 0.5%. This is a large (100-fold) increase, but still is only 1 out of every 200 persons, and, given the low starting point, should be achievable.
- Establish compulsory service programs for young adults in at least 50% of the world's nations.
- Achieve the annual participation of at least 50% of all people worldwide to the accomplishment of long term goals through contributions of their money and/or labor.

ACTIONS WE CAN ALL TAKE

- Make financial contributions, even small ones, to nonprofit organizations working toward worthwhile long term goals.
- Contribute time to activities of the non-profits.
- Consider a career working toward accomplishment of worthwhile long term goals.
- Devote more time to the goals when in retirement.
- Vote for political leaders who will establish worthy long term goals and strive to attain them.

Chapter 10

Concluding Remarks

I THINK THE MOST IMPORTANT idea in this book is that humanity needs to set and accomplish long term goals. My own particular list of goals is secondary. Setting long term goals will mark the end of human infancy and the beginning of our childhood. It will fix a rudder to the human ship. I hope this slim book will catalyze thought, discussion, and appreciation for long term goals. More than anything else, that was my purpose.

In devising my suggested goals, I prioritized reduction of human suffering over expansion of human pleasure. In my experience, people don't need encouragement to seek pleasure, but they are much less attuned to the suffering of others. Reducing suffering requires goals, plans, and great effort.

Science has leapt so far ahead of other explanations for human progress that, although I believe we absolutely must continue to invest in science, I think much more effort needs to be made to improve other areas of our societies, such as governments, international relations, and economics. There are perhaps 10 million researchers in the world today working full time to advance scientific knowledge (1), but probably fewer than 100 working full time to establish an effective world government. We need these other areas of our societies to at least begin to catch up to science.

I have confidence that the goals described in this book are worthwhile, but I have less confidence in the targets listed at the end of each chapter. It was difficult for me to gauge the correct level of ambition for the targets, especially since most of the goals were outside of my expertise. Even if the goals themselves endure, probably many of the targets will need to be tuned by experts.

My suggested goals may strike many readers as overly ambitious and impossible to achieve, but we need to remember that these are ten generation goals. If our thinking is restricted to a few years or a few decades, then yes, my goals are unrealistic. But if our time frame is centuries, then I believe the goals are eminently achievable.

Goals longer term than 300 years—say, 3,000 or even 30,000-year goals—can certainly be envisioned. I think that such really long term goals would be considerably harder to formulate. It's difficult enough to project human progress 300 years into the future, much less 30,000. Still, such really long term goals may be useful. I leave it to others to explore this possibility.

Some critics of this book will say that while there is clearly a need for long term goals, my particular list of goals is pedestrian. To such critics, I say fine; I challenge you to devise better lists. My suggested goals are certainly not the only ones that should be considered, nor do I claim that my list is the best one. People with different experiences and perspectives, especially, perhaps, those from other parts of the world, will likely devise quite different lists. This is all good. The more people working on this task, the better.

- - -

To teachers and mentors of young people in the age range of about 16–20, I recommend three assignments that I think will both benefit the young people and raise awareness of long term goals.

The first assignment is for the students to make their own lists of long term goals for humanity (perhaps five goals per student). I think the students' lists would be fascinating. Certainly, many of the students' goals would be impractical or possibly even harmful, but preparing and discussing the lists would encourage the students to begin thinking about long term goals.

The second assignment is for the students to prepare lists of criteria by which people's old-age pensions would be set (Chapter 7). What do the students think is a "good life"? My own list of preferred criteria is presented in Appendix 1.

The third assignment is for students to prepare prospectively their own life summaries of no more than 1,500 words (also Chapter 7). This assignment would help students think about what they would like to achieve in their own lives. These predictive autobiographies would perhaps be too personal to be discussed in a class, but I think the exercise would still be of substantial benefit. In Appendix 2, I provide (retrospectively) my own life summary.

- - -

I believe that humanity's greatest resource is not fertile agricultural land, or temperate climate, or rich deposits of minerals, or even knowledge, but rather the time and energy of people. While small groups of influential individuals may initiate and organize, I don't think that long term goals will ever be achieved without contributions from entire populations.

I hope that the lists of actions at the end of each chapter are good starting points for individuals. These actions don't require great personal sacrifice. You do not need to be among the most talented and highly educated to make meaningful contributions. Many readers will have noticed that voting for leaders who will support the long term goals is included in multiple lists at the ends of the chapters. Voting is a relatively simple action that doesn't usually require extraordinary effort, yet is critically important. I feel that voting is among the highest duties of citizens in free nations.

Young people especially in their teens and early 20s are often idealistic and full of energy. They wonder why humanity's big problems haven't already been solved. They think that when they get older, they will easily correct these problems. Then, as they mature, they come to realize that the big problems of humanity are *really* tough to solve. Consequently, they often become discouraged and abandon all effort. But I say that giving up is a huge mistake. Yes, the great problems of humanity are really difficult to solve, and yes, no one can solve them alone, but that doesn't mean that they can't be solved.

> Nobody makes a greater mistake than he who does nothing because he could do only a little.
>
> Attributed to Edmund Burke,
> member of the English Parliament
> (1729–1797)

Human happiness does not arise solely from pleasure, but also from engaging in worthwhile tasks, especially, perhaps, when such tasks require considerable effort and sacrifice (2–5). Working toward worthwhile long term goals will produce much health, happiness, and satisfaction. Consider the following story that I heard from a pastor at a church service years ago:

> A man in Europe in the Middle Ages was walking down a hill away from a town. Along the way he met three men pushing carts of stones up the hill. He asked the first man what he was doing. The man replied: "I have to push these rocks up the hill. It must be the worst job in the world. My boss is a slave driver, they hardly pay me anything, and my back is killing me." The

> traveler asked the second man what he was doing. He said: "Well, this isn't the best job around, but it's all I could find, and I need the job to support my wife and children." The third man the traveler met had a smile on his face. When asked what he was doing, he replied: "I'm building a cathedral."

Although I'm not a particularly optimistic person, I also reject blanket pessimism. Humanity certainly has no shortage of horrors, disasters, and tragedies on our ledger, but we can also point to a long string of successes. Over the last few centuries, infant, child, and maternal mortality have been greatly reduced; slavery has been eliminated; violence has diminished (6); a substantial fraction of people now enjoys basic human rights.

I don't know if long term goals will ever be accomplished or even set, but I am certain that humanity has the ability to do so. I challenge future generations to vanquish despair and seize a brighter future.

Appendix 1

My Criteria for Evaluating Human Lives

AS DESCRIBED IN CHAPTER 7, if government old-age pensions and/or healthcare benefits are to be based on performance during life, then criteria need to be developed to evaluate human lives. There are obviously a huge number of options. I suggested in Chapter 7 three criteria as a starting point: employment, criminal activity, and voting. This appendix provides an expanded view of my favored criteria. I express my preferences in the form of a series of questions.

I don't expect people to be saints. I also try not to be a prude. The purpose of the criteria is not to demand unrealistic perfection, but rather to fairly evaluate people relative to each other, and especially to encourage better behavior. Even small improvements will help.

There are many virtues, like humility and courtesy, that are not included in my list of criteria. There are also many human faults, like lying and cheating, that are likewise not included. Such virtues and faults are excluded not because they are unimportant, but rather because they are too difficult to monitor.

One theme of my criteria is that everyone has a responsibility to contribute positively to human societies. We all derive tremendous benefit from living within our modern societies. We receive countless assists in our lives from enormous numbers of people with a dizzying array of special experience, knowledge, and skills. Consider the multitude of people involved in bringing an appliance like a refrigerator into our homes. These include everyone from the miners who extract the raw materials, to the design engineers, to the manufacturers, to those who transport the finished products, to those who generate the electricity to run the appliance, and many, many others. Because we all receive such great benefit from the efforts of others, I believe that it is only just that we also make our own contributions to society in return.

A second theme is that those who are fortunate in life to be healthier and more talented than most have a responsibility to contribute more to society than the average person. I'm convinced that much of what society generally considers success in life is due to factors beyond our control, such as family and genome. The healthier and the more talented the person, the more that should be expected from them.

A third theme is that the decisions we make and the actions we take in life are ultimately most important in determining what type of life we lead. Each of us makes many decisions that strongly affect others—decisions about questions such as "Should I work hard at my job?" or "Should I cheat on my taxes?" Many with poor health and little talent make excellent decisions and are assets to society, and many who are exceptionally healthy and talented make bad decisions and are liabilities.

- - -

1. Has the person worked during their adult years, and were there gaps? Our societies need every person to contribute with their labor to the best of their abilities.
2. Has the person obeyed the law, or have they engaged in criminal activity? All government criminal records, from minor to major crimes, should be retained.
3. Did the person vote in elections? I think our freedom is absolutely dependent upon voting in elections.
4. If the person had children, what are the records of the children? In my opinion, it is essential that parents put considerable effort into and take responsibility for raising their children.
5. If the person married one or more times, how long did the marriages last? I feel that dedication and loyalty to a spouse is important.
6. Has the person ever been dismissed from a job, and if so, why? Many lose jobs through no fault of their

own, but many others lose their jobs for cause. Has the person held relatively few or many jobs throughout their lifetime? I think that loyalty to employers can be overemphasized, but as a former employer, I've also found that people who change jobs frequently usually do not make good employees. Employers desperately want good employees and will strive to retain them.

7. Did the person serve as a police officer or firefighter or with the military? Anyone who places themselves at personal risk to protect and defend others, in my opinion, deserves honor and reward.
8. What are the person's major accomplishments? Effort itself is desirable and laudable, but results are what ultimately count. Accomplishments are certainly difficult to quantify. How, for example, does one compare the composition of a hit song to serving as town mayor? Still, I think lists of accomplishments will be helpful in evaluation. Lists should be verified wherever possible.
9. Did the person's accomplishments match their health and abilities? This question is even more difficult to evaluate than the previous one, but I think it is critically important. Of course, assessing health and talent is difficult, but computer programs can probably extract a reasonable measure of health from electronic health records, and problem-solving ability can be roughly gauged by using school achievement tests and educational attainment. By then comparing accomplishments

to health and talent, we can obtain a rough idea of whether the person over or underachieved in life. As recordkeeping and storage capabilities improve, it should be possible over time to gradually refine this approach.

10. How much did the person contribute through both money and time to help achieve worthwhile long term goals? Records of these actions should be collected and retained.

- - -

Some of these criteria can't effectively apply to people living in totalitarian states. In a free nation, a person serving as a police officer, for example, should, I think, be honored, but a police officer doing the dirty work of a tyrant is dishonorable.

What about the awful person who, say, commits crimes, but who also writes popular novels or achieves major scientific breakthroughs? No one is perfect. For everyone, I think we need to balance the good against the bad, the achievements against the mistakes.

I also think it's important for all people to understand that it is never too late to turn their lives around. If the criteria are set so that mistakes early in life mean that people can never earn good scores, then there is no motivation for rededication. Special credit should be given to those who recover later in life.

Appendix 2

Life Summary for James Lewis Weber

I WAS FORTUNATE TO BE born into a stable home with devoted and loving parents. I was also fortunate to have lived in a democratic and affluent nation. My family was middle class: not wealthy, but also not poor. My mother was a nurse; my father, an accountant. I was healthy and fairly energetic most of my life.

I grew up and went to school in a middle-income neighborhood adjacent to the city of Milwaukee, state of Wisconsin, United States. I was always a talented student. Because I didn't have to take jobs to support myself or my family during grade or high school, I was able to concentrate on my studies and also participate in extracurricular activities such as music and sports.

After high school, at age 17, I entered the US Military Academy at West Point, a training college for Army officers. I very soon realized that a military career was not for me. I resigned from the academy after only about three weeks. In hindsight, I don't regret this decision, but still, especially when I was younger, I considered my resignation a huge failure. My experience at West Point was an important event in my life. It made me especially determined not to fail again.

After West Point, I enrolled at the University of Wisconsin–Milwaukee, where I was a full-time student majoring in chemistry. During this period, I also held a job for 18 hours per week as a proofreader for the *Milwaukee Journal* newspaper. I lived in my parents' home.

After two years at the University of Wisconsin–Milwaukee, I transferred to the main campus of the University of Wisconsin in Madison and continued my undergraduate studies in chemistry and biology. I took difficult and sometimes overly ambitious courses. Nevertheless, I graduated with honors as a chemistry major in 1974.

I married before my last semester at Madison. My wife and I have remained married for 52 years and counting. We started life together with few possessions and almost no money.

After Madison, I entered graduate school at the University of California at Berkeley, where I studied biochemistry. The biochemistry department at Berkeley was top notch. I learned a great deal at Berkeley. I graduated with my PhD in 1980.

After graduate school, I returned to the University of Wisconsin–Madison where I worked as a postdoctoral research fellow for two years. While in California, I greatly missed the midwestern part of the US and wanted to return. I coauthored a few scientific publications while a graduate student and postdoc, but nothing particularly noteworthy.

I next joined the US Army as a captain in the Medical Service Corps. After some basic training in San Antonio, Texas, I worked for four years (1982–1986) on malaria vaccines at the Walter Reed Army Medical Center in Washington, DC. I was relatively productive at Walter Reed, authoring a number of scientific publications and contributing to a team that made progress toward the difficult malaria vaccine. I did little soldiering while in the Army, but instead spent almost all of my time in the laboratory. I had always worked hard, but until then I was relatively undisciplined. The Army improved my personal discipline considerably.

After Washington, my wife and I moved to Marshfield, Wisconsin, where I took a position as a research scientist at the nonprofit Marshfield Clinic Research Institute. Marshfield was a small city (population approximately 20,000) located in a rural part of central Wisconsin. In 1985, 15 months before we left Washington, DC, our daughter was born, and just two months after we arrived in Marshfield, in 1986, our son was born. Our financial situation slowly improved.

I worked as a research scientist in Marshfield for about 18 years (1986–2004). I was productive during this period. Over my entire research career, I authored or coauthored over 200 publications in scientific journals. I competed successfully for a number of large research grants. At its peak, my research lab employed about 35 scientists and engineers. My biggest research achievements were the discovery of DNA short tandem repeat polymorphisms and application of these polymorphisms to map disease genes and to build some of the first chromosomal maps as part of the Human Genome Project.

My wife and I devoted much time during those years to raising our children. We tried hard to be good parents.

After my time as a nonprofit research scientist, I founded a for-profit clinical DNA testing company in Marshfield. I served as the head of this company for 18 years (2004–2021). During this period, the company grew from about 6 to 220 employees. We provided clinical DNA tests for approximately 500,000 patients. Leading a company was difficult and stressful, but I enjoyed employing people and helping patients. In 2021, at the age of 69, I sold the company and semiretired.

In my semiretirement, I wrote two books: *Suggestions for Long Term Goals for Humanity* (2026) and *Genomes and the Gene Pool* (2026).

I exercised nearly every day of my adult life. I'm convinced that the exercise helped me maintain good health.

I voted in nearly every election.

I believe I was a kind and generous person, but I was never outgoing. I think most people considered me serious and reserved, even cold or unfriendly at times.

Although there are certainly many who worked harder, I did work hard throughout my life, several times to the point that I became ill. I do not think that hard work is inconsistent with happiness.

Besides work, I enjoyed many other aspects of life, including family, music, reading, sports, and being outdoors in nature.

With a good portion of the wealth received from the sale of the testing company, I started a local charitable foundation. I also purchased some land and donated it to the state of Wisconsin to expand a wildlife reserve. My lifestyle changed little after the sale.

Considering all the horrors of war, famine, pestilence, and other disasters that *many* other people have experienced throughout human history, I lived a relatively charmed life. There were certainly obstacles and challenges, but no such disasters.

In my life, I made many mistakes and did many things of which I'm not proud, but my goal was always to try to make the world a better place. Literally, when I awoke each day, I set out in my imperfect and limited fashion to do the best I could to advance humanity. The way I liked to express it was "What counts in life is not what you do *to* others, but rather what you do *for* others." Certainly not a novel concept, but still a good way to live.

References

CHAPTER 1: LONG TERM GOALS FOR HUMANITY

1-1. Paris Climate Agreement of 2015 (ttps://unfccc.int/sites/default/files/resource/english_paris_agreement.pdf).

1-2. How science and cooperation closed the ozone hole. Nature. 641:559-560, 2025.

1-3. United Nations Developmental Goals (https://sdgs.un.org/goals).

1-4. Global Sustainable Development Report 2025 from the United Nations (https://unstats.un.org/sdgs/report/2025/The-Sustainable-Development-Goals-Report-2025.pdf).

CHAPTER 2: CONTINUE THE HUMAN SPECIES

2-1. Fortney R. Life: A Natural History of the First Four Billion Years of Life on Earth. Vintage Books. 1997.

2-2. Gee HA. (Very) Short History of Life on Earth. St. Martin's Press. 2021.

2-3. Roberts A, editor. Evolution, the Human Story. Penguin Random House. Second Edition. 2018.

2-4. Higham T. The World Before Us, The New Science Behind Our Human Origins. Yale University Press. 2021.

2-5. Reich D. Who We Are and How We Got Here, Ancient DNA and the New Science of the Human Past. Pantheon Books. 2018.

2-6. Alvarez LW, Alvarez W, Asaro F, Michel HV. Extraterrestrial cause for the cretaceous-tertiary extinction. Science. 208:1095-1108, 1980. PubMed ID: 17783054.

2-7. Becker L. Repeated blows. Scientific American. 286(3):76-83, 2002. PubMed ID: 11857903.

2-8. Fuller E. The Passenger Pigeon. Princeton University Press. 2015.

2-9. Ord T. The Precipice, Existential Risk and the Future of Humanity. Hachette Books. 2020. (See p.167 for a table of the probabilities of risks.)

2-10. Posner RA. Catastrophe, Risk and Response. Oxford University Press. 2004.

2-11. Bostrom N, Cirkovic MM, Editors. Global Catastrophic Risks. Oxford University Press. 2008.

2-12. Walsh B. End Times, A Brief Guide to the End of the World. Hachette Books. 2019.

2-13. Ananthaswamy A. How close is AI to human-level intelligence? Nature. 636(8041):22-25, 2024. PubMed ID: 39627530.

2-14. Bengio Y et al. Managing extreme AI risks amid rapid progress. Science. 384(6698):842-845, 2024. PubMed ID: 38768279.

2-15. Kurzweil R. The Singularity is Nearer: When We Merge with AI. Viking. 2024.

2-16. Yudkowsky E, Soares N. If Anyone Builds It, Everyone Dies; Why Superhuman AI Would Kill Us All. Little, Brown and Company. 2025.
2-17. Center for AI Safety, 2023 (https://www.safe.ai/work/statement-on-ai-risk).
2-18. Collision of Comet Shoemaker-Levy 9 with Jupiter (https://en.wikipedia.org/wiki/Comet_Shoemaker%E2%80%93Levy_9#Impacts).
2-19. Chapman CR, Morrison D. Impacts on the earth by asteroids and comets: assessing the hazard. Nature. 367:33-40, 1994.
2-20. Napier R. Hazards from Comets and Asteroids. In Bostrom N, Cirkovic MM, Editors. Global Catastrophic Risks. Oxford University Press. pp. 222-237, 2008.
2-21. Andrews RG. Planet protector. Science. 390(6778):1094-1099, 2025. PubMed ID: 41379959.
2-22. Seligman DZ et al. Two distinct populations of dark comets delineated by orbits and sizes. Proceedings of the National Academy of Sciences USA. 121(51):e2406424121, 2024. PubMed ID: 39652762.
2-23. The 'Oumuamua ISSI Team. The natural history of 'Oumuamua. Nature Astronomy 3:594-602, 2019.
2-24. The DART Space Mission (https://www.space.com/dart-asteroid-mission).
2-25. Adam CD. DART's data verify its smashing success at deflecting asteroid moon Dimorphos. Nature. 616:437-438, 2023. PubMed ID: 37076732.
2-26. Herbert F. The White Plague. G.P. Putnam's Sons. 1982.
2-27. Berns KI et al. Public health and biosecurity. Adaptations of avian flu virus are a cause for concern. Science. 335:660-661, 2012. PubMed ID: 22294736.

2-28. Kaiser J. Making trouble. Science 378:242-245, 2022. PubMed ID: 36264794.
2-29. Msemburi W et al. The WHO estimates of excess mortality associated with the COVID-19 pandemic. Nature. 613:130-137, 2023. PubMed ID: 36517599.
2-30. COVID-19 Excess Mortality Collaborators. Estimating excess mortality due to the COVID-19 pandemic: a systematic analysis of COVID-19-related mortality, 2020-21. Lancet. 399:1513-1536, 2022. PubMed ID: 35279232.
2-31. Adamala KP et al. Confronting risks of mirror life. Science. 386(6728):1351-1353, 2024. PubMed ID: 39666824.
2-32. Bindeman IN. The secrets of supervolcanoes. Scientific American. 294(6):36-43, 2006. PubMed ID: 16711358.
2-33. Rampino MR. Super-volcanism and other geophysical processes of catastrophic import. In Bostrom N, Cirkovic MM, Editors. Global Catastrophic Risks. Oxford University Press. pp. 205-221, 2008.
2-34. Cassidy M, Mani L. Prepare now for big eruptions. Nature. 608:469-471, 2022. PubMed ID: 35978116.
2-35. Turco RP et al. Climate and smoke: an appraisal of nuclear winter. Science. 247:166-176, 1990. PubMed ID: 11538069.
2-36. Vilhelmsson A, Baum SD. Public health and nuclear winter: addressing a catastrophic threat. Journal of Public Health Policy. 44(3):360-369, 2023. PubMed ID: 37322225.
2-37. Weber JL. Genomes and the Gene Pool. Ten Generations Press. 2026.

2-38. Olson S. Mapping Human History. Houghton Mifflin. 2002.

2-39. Records of the General Conference, United Nations Educational, Scientific and Cultural Organization (UNESCO) 29th session, Paris, 21 October to 12 November 1997, v. 1: Resolutions (https://en.unesco.org/about-us/legal-affairs/declaration-responsibilities-present-generations-towards-future-generations).

CHAPTER 3: CONTINUE OTHER SPECIES

3-1. Hannah M. Extinctions. Cambridge University Press. 2021.

3-2. Barnosky AD et al. Has the Earth's sixth mass extinction already arrived? Nature. 471:51-57, 2011. PubMed ID: 21368823.

3-3. Diamond JM. The present, past and future of human-caused extinctions. Philosophical Transactions of the Royal Society of London B Biological Sciences. 325(1228):469-476, 1989. PubMed ID: 2574887.

3-4. Cohen JE. How Many People Can the Earth Support? W. W. Norton and Company. 1995.

3-5. Keinan A, Clark AG. Recent explosive human population growth has resulted in an excess of rare genetic variants. Science. 336(6082):740-743, 2012. PubMed ID: 22582263.

3-6. Worldometer Population Estimates (https://www.worldometers.info/world-population/world-population-by-year/).

3-7. Vitousek PM, Mooney HA, Lubchenco J, Melillo JM. Human Domination of Earth's Ecosystems. Science. 277(5325): 494-499, 1997.

3-8. Díaz S et al. (eds.). Summary for policymakers of the global assessment report on biodiversity and ecosystem services of the Intergovernmental Science-Policy Platform on Biodiversity and Ecosystem Services (IPBES). IPBES secretariat, Bonn, Germany (https://doi.org/10.5281/zenodo.3553579). 2019.

3-9. Williams DR, Rondinini C, Tilman D. Global protected areas seem insufficient to safeguard half of the world's mammals from human-induced extinction. Proceedings of the National Academy of Sciences USA. 119(24):e2200118119, 2022. PubMed ID: 35666869.

3-10. Rosenberg KV et al. Decline of the North American avifauna. Science. 366(6461):120-124, 2019. PubMed ID: 31604313.

3-11. Stuart SN et al. Status and trends of amphibian declines and extinctions worldwide. Science. 306(5702):1783-1786, 2004. PubMed ID: 15486254.

3-12. Cox N et al. A global reptile assessment highlights shared conservation needs of tetrapods. Nature. 605(7909):285-290, 2022. PubMed ID: 35477765.

3-13. Almond REA, Grooten M, Juffe Bignoli D, Petersen T (eds). Living Planet Report 2022 – Building a nature positive society. World Wildlife Federation, Gland, Switzerland (https://livingplanet.panda.org/).

3-14. Living Planet Index, 2024 (https://www.livingplanetindex.org/).

3-15. Wagner DL et al. Insect decline in the Anthropocene: Death by a thousand cuts. Proceedings of the National Academy of Sciences USA. 118(2):e2023989118, 2021. PubMed ID: 33431573.

3-16. Edwards CB et al. Rapid butterfly declines across the United States during the 21st century. Science. 387(6738):1090-1094, 2025. PubMed ID: 40048533.

3-17. Walsh W, Figueroa LL. Data deficiency, taxonomic bias, and economic interests curtail insect and arachnid conservation in the United States. Proceedings of the National Academy of Sciences USA. 123(10):e2522779123, 2026. PubMed ID: 41770926.

3-18. Sousa R. A brief global agenda for advancing the study of molluscs. Frontiers in Ecology and Evolution. 12:1176380, 2024.

3-19. Runyan C, D'Odorico P. Global Deforestation. Cambridge University Press. 2016.

3-20. Gibbs WW. On the termination of species. Scientific American 285(5):40-49, 2001. PubMed ID: 11681175.

3-21. Finn C, Grattarola F, Pincheira-Donoso D. More losers than winners: investigating Anthropocene defaunation through the diversity of population trends. Biological Reviews of the Cambridge Philosophical Society. 98(5):1732-1748, 2023. PubMed ID: 37189305.

3-22. Ceballos G, Ehrlich PR, Dirzo R. Biological annihilation via the ongoing sixth mass extinction signaled by vertebrate population losses and declines. Proceedings of the National Academy of Sciences USA. 114(30):E6089-E6096, 2017. PubMed ID: 28696295.

3-23. Wilson EO. Half-Earth. W. W. Norton. 2016.

3-24. IUCN. 2022. The IUCN Red List of Threatened Species. Version 2022-2. (https://www.iucnredlist.org).

3-25. Costanza R et al. The value of the world's ecosystem services and natural capital. Nature. 387: 253–260, 1997.

3-26. Balmford A et al. Economic reasons for conserving wild nature. Science. 297(5583):950-953, 2002. PubMed ID: 12169718.

3-27. World Economic Forum. Nature risk rising: why the crisis engulfing nature matters for business and the economy (https://www3.weforum.org/docs/WEF_New_Nature_Economy_Report_2020.pdf).

3-28. Díaz S et al. Pervasive human-driven decline of life on Earth points to the need for transformative change. Science. 366(6471):eaax3100, 2019. PubMed ID: 31831642.

3-29. Wang N et al. Genomic conservation of crop wild relatives: A case study of citrus. PLoS Genetics. 19(6):e1010811, 2023. PubMed ID: 37339133.

3-30. Artaxo P. Amazon deforestation implications in local/regional climate change. Proceedings of the National Academy of Sciences USA. 120(50):e2317456120, 2023. PubMed ID: 38032950.

3-31. Lang C, VanCeylon J, Ando AW. Distribution of capitalized benefits from land conservation. Proceedings of the National Academy of Sciences USA. 120(18):e2215262120, 2023. PubMed ID: 37094159.

3-32. Rodger JG et al. Widespread vulnerability of flowering plant seed production to pollinator declines. Science Advances. 7(42):eabd3524, 2021. PubMed ID: 34644118.

3-33. Cornelisse T et al. Elevated extinction risk in over one-fifth of native North American pollinators. Proceedings of the National Academy of Science USA. 122(14):e2418742122, 2025. PubMed ID: 40127257.

3-34. Neugarten RA et al. Mapping the planet's critical areas for biodiversity and nature's contributions to people. Nature Communications. 15(1):261, 2024. PubMed ID: 38199986.

3-35. Visconti P et al. Protected area targets post-2020. Science 364(6437):239-241, 2019. PubMed ID: 30975769.

3-36. Michaelsen P, Sundström A, Jagers SC. Mass support for conserving 30% of the Earth by 2030: Experimental evidence from five continents. Proceedings of the National Academy of Science USA. 122(35):e2503355122, 2025. PubMed ID: 40857320.

3-37. Jones KR et al. One-third of global protected land is under intense human pressure. Science. 360(6390):788-791, 2018. PubMed ID: 29773750.

3-38. Brodie JF et al. A well-connected Earth: The science and conservation of organismal movement. Science. 388(6745):eadn2225, 2025. PubMed ID: 40273266.

3-39. Wang S, Yoder AD. Synergies between speciation and conservation science yield novel insights for mitigating the biodiversity crisis of the Anthropocene. Proceedings of the National Academy of Science USA. 122(31):e2500713122, 2025. PubMed ID: 40720642.

3-40. Zou Y et al. Fragmentation increased in over half of global forests from 2000 to 2020. Science. 389(6765):1151-1156, 2025. PubMed ID: 40934301.

3-41. Potapov P et al. Global maps of cropland extent and change show accelerated cropland expansion in the twenty-first century. Nature Food. 3(1):19-28, 2022. PMID: 37118483.

3-42. Strassburg BBN et al. Global priority areas for ecosystem restoration. Nature. 586(7831):724-729, 2020. PubMed ID: 33057198.
3-43. Jung M et al. Areas of global importance for conserving terrestrial biodiversity, carbon and water. Nature Ecology and Evolution. 5(11):1499-1509, 2021. PubMed ID: 34429536.
3-44. United Nations World Population Prospects (https://population.un.org/wpp/).
3-45. Metz T et al. Biodiversity resilience in a tropical rainforest. Nature. 652(8112):1232-1239, 2026. PubMed ID: 41951739.
3-46. International Crane Foundation (https://savingcranes.org/).
3-47. US Fish and Wildlife Service (https://www.fws.gov/press-release/2023-05/2023-wintering-whooping-crane-count).
3-48. Anderson MG et al. A resilient and connected network of sites to sustain biodiversity under a changing climate. Proceedings of the National Academy of Sciences USA. 120(7):e2204434119, 2023. PubMed ID: 36745800.
3-49. Allan JR et al. The minimum land area requiring conservation attention to safeguard biodiversity. Science 376(6597):1094-1101, 2022. PubMed ID: 35653463.

CHAPTER 4: ESTABLISH A DEMOCRATIC WORLD GOVERNMENT

4-1. UN Universal Declaration of Human Rights (https://www.un.org/en/about-us/universal-declaration-of-human-rights) 1948.
4-2. Kant I. Perpetual Peace. The Bobbs-Merrill Company Inc. 1795.

4-3. Streit CK. Union Now. Jonathan Cape Ltd. 1939.
4-4. Reves E. The Anatomy of Peace. Harper and Brothers. 1945.
4-5. Hudgens TA. Let's Abolish War (Not a New Idea). BILR Corporation. ISBN 0-937177-00-8. 1986.
4-6. Leinen J, Bummel A. A World Parliament, Governance and Democracy in the 21st Century. Democracy without Borders, Berlin. 2018
4-7. World Federalist Movement/Institute for Global Policy (https://www.wfm-igp.org/).
4-8. Democracy Without Borders (https://www.democracywithoutborders.org/about/).
4-9. Young World Federalists (https://www.ywf.world/).
4-10. Streit Council for a Union of Democracies (https://www.streitcouncil.org/history).
4-11. White M. The Great Big Book of Horrible Things. W.W. Norton and Company. 2011.
4-12. Diamond J. The World Until Yesterday; What Can We Learn from Traditional Societies. Viking. 2012.
4-13. Weber JL. Genomes and the Gene Pool. Ten Generations Press. 2026.
4-14. SEAMLESS Communication Team. Joint speech and text machine translation for up to 100 languages. Nature. 637(8046):587-593, 2025. PubMed ID: 39815098.
4-15. Montgomery SL. Does Science Need a Global Language? University of Chicago Press. 2013.
4-16. Gordin MD. Scientific Babel, The Language of Science from the Fall of Latin to the Rise of English. University of Chicago Press. 2015.
4-17. Salomone R. The Rise of English. Oxford University Press. 2022.

4-18. Ahamed MF. Difficulties and Challenges in Teaching English as the Second and Foreign Language. The Asian Conference on Language Learning 2013 Official Conference Proceedings. (https://papers.iafor.org/wp-content/uploads/papers/acll2013/ACLL2013_0140.pdf) 2013.
4-19. Paulesu E et al. A cultural effect on brain function. Nature Neuroscience. 3(1):91-96, 2000. PubMed ID: 10607401.
4-20. Paulesu E et al. Dyslexia: cultural diversity and biological unity. Science. 291(5511):2165-2167, 2001. PubMed ID: 11251124.
4-21. Ho CS, Fong KM. Do Chinese dyslexic children have difficulties learning English as a second language? Journal of Psycholinguist Research. 34(6):603-618, 2005. PubMed ID: 16341916.
4-22. Pinker S. The Better Angels of our Nature; Why Violence Has Declined. Viking Press. 2011.
4-23. Our World in Data Democracy Index (https://ourworldindata.org/grapher/democracy-index-eiu?time=latest).
4-24. Economist Intelligence Unit Democracy Index 2023 (https://www.eiu.com/n/campaigns/democracy-index-2023/).

CHAPTER 5: MANAGE THE HUMAN GENE POOL

5-1. Weber JL. Genomes and the Gene Pool. Ten Generations Press. 2026.
5-2. Darwin C, Wallace AR. On the Tendency of Species to Form Varieties; and on the Perpetuation of Varieties and Species by Natural Means of Selection. Zoological Journal of the Linnean Society. 3(9):45–62, 1858.

5-3. Darwin C. Origin of the Species. Penguin Classics. 1st Edition, 1859.
5-4. Oeppen J, Vaupel JW. Demography. Broken limits to life expectancy. Science. 296(5570):1029-1031, 2002. PubMed ID: 12004104.
5-5 Johnson S. Extra Life, The Astonishing Story of How We Doubled Our Lifespan. Viking. 2023.
5-6. Dattani S, Spooner F, Ritchie H, Roser M. (2023) - "Child and Infant Mortality" Published online at OurWorldinData.org. 2023. (https://ourworldindata.org/child-mortality).
5-7. Reilly PR. Eugenics and Involuntary Sterilization: 1907-2015. Annual Review of Genomics and Human Genetics. 16:351-368, 2015. PubMed ID: 26322647.
5-8. Armstrong C. Thousands of women sterilised in Sweden without consent. British Medical Journal. 315(7108):563, 1997. PubMed ID: 9302943.
5-9. Müller-Hill B. Murderous Science (Tödliche Wissenschaft). Cold Spring Harbor Press. 1984 (German original); 1998 (English translation).

CHAPTER 6: GROW KNOWLEDGE

6-1. Darwin C. Origin of the Species. Penguin Classics. 1st Edition, 1859.
6-2. Ritchie H, Rosado P, Roser M. Crop Yields. Published online at OurWorldInData.org, 2022 (https://ourworldindata.org/crop-yields).
6-3. Roser M, Ortiz-Ospina E, Ritchie H. Life Expectancy. Published online at OurWorldInData.org. (https://ourworldindata.org/life-expectancy) 2013.
6-4. Clark G. A Farewell to Alms. Princeton University Press. 2007. (see Figure 1.1 on page 2)
6-5. Susskind D. Growth, A History and a Reckoning. The Belknap Press of Harvard University Press. 2024.

6-6. Bush V. Science, the Endless Frontier. Princeton University Press. 1945.

6-7. Press WH. What's so special about science (and how much should we spend on it?). Science. 342(6160):817-822, 2013. PubMed ID: 24244958.

6-8. Salter AJ, Martin BR. The economic benefit of publicly funded basic research: a critical review. Research Policy. 30:509-532, 2001.

6-9. The-Impact-of-Genomics-on-the-US-Economy (http://unitedformedicalresearch.org/wp-content/uploads/2013/06/) 2013.

6-10. Wu R, Esposito C, Evans J. Shifting power asymmetries in scientific teams reveal China's rising leadership in global science. Proceedings of the National Academy of Sciences USA. 122(44):e241489312, 2025. PubMed ID: 41150718.

6-11. Bolick J et al. Planet of the APCs: A decade of progress and setbacks in open access. Journal of Librarianship and Scholarly Communication. 12(2): eP18319, 2024.

6-12. Alberts B. The new Egypt. Science. 332(6029):513, 2011. PubMed ID: 21527679.

6-13. Mungiu-Pippidi A. Good governance powers innovation. Nature. 518(7539):295-297, 2015. PubMed ID: 25693546.

6-14. Miller JD, Scott EC, Okamoto S. Public acceptance of evolution. Science. 313(5788):765-766, 2006. PubMed ID: 16902112.

6-15. Hameed S. Science and religion. Bracing for Islamic creationism. Science. 322(5908):1637-1638, 2008. PubMed ID: 19074331.

6-16. Organization for Economic Cooperation and Development Gross Domestic Spending on R&D (https://data.oecd.org/rd/gross-domestic-spending-on-r-d.htm).

CHAPTER 7: TEMPER GREED

7-1. Plato. Republic. ~360 BCE.
7-2. Smith A. The Wealth of Nations. 1776.
7-3. Diamond, J. Guns, Germs and Steel. W. W. Norton & Company. 1999.
7-4. Lynn R, VanHanen T. IQ and the Wealth of Nations. Praeger Greenwood. 2002.
7-5. Fukuyama F. The Origins of Political Order. Farrar, Straus and Giroux. 2011.
7-6. Henrich J. The WIERDest People in the World. Farrar, Straus and Giroux. 2020.
7-7. Mauro P. Corruption and Growth. The Quarterly Journal of Economics. 110(3):681-712, 1995.
7-8. Christos P et al. Corruption Perception Index (CPI), as an Index of Economic Growth for European Countries. Theoretical Economics Letters. 8:524-537, 2018.
7-9. Gründler K, Potrafke N. Corruption and economic growth: new empirical evidence. European Journal of Political Economy. 60:101810, 2019.
7-10. Mungiu-Pippidi A. Good governance powers innovation. Nature. 518(7539):295-297, 2015. PubMed ID: 25693546.
7-11. Rawls J. A Theory of Justice. The Belknap Press of Harvard University Press. 1971. (p. 6)
7-12. Rau EG, Stokes S. Income inequality and the erosion of democracy in the twenty-first century. Proceedings of the National Academy of Science USA. 122(1):e2422543121, 2025. PubMed ID: 39793070.

7-13. Wilkinson R, Pickett K. The Spirit Level, Why Greater Equality Makes Societies Stronger. Bloomsbury Press. 2009.

7-14. Kaiser C, Oswald AJ. Inequality, well-being, and the problem of the unknown reporting function. Proceedings of the National Academy of Sciences USA. 119(50):e2217750119, 2022. PubMed ID: 36472960.

7-15. Dwyer RJ, Dunn EW. Wealth redistribution promotes happiness. Proceedings of the National Academy of Sciences USA. 119(46):e2211123119, 2022. PubMed ID: 36343268.

7-16. Wade L. Birth of the moralizing gods. Science. 349(6251):918-922, 2015. PubMed ID: 26315417.

7-17. Purzycki BG et al. Moralistic gods, supernatural punishment and the expansion of human sociality. Nature. 530(7590):327-330, 2016. PubMed ID: 26863190.

7-18. Saez E, Zucman G. The Triumph of Injustice, How the Rich Dodge Taxes and How to Make Them Pay. W.W. Norton and Company. 2019.

7-19. Robeyns I. Limitarianism, The Case Against Extreme Wealth. Astra House. 2024.

7-20. Piketty T. Capital in the Twenty-First Century. Belknap Press. 2013.

7-21. Carnegie A. The Gospel of Wealth (https://www.carnegie.org/about/our-history/gospelofwealth/). 1889.

7-22. Piketty T, Saez E. Inequality in the long run. Science. 344(6186):838-843, 2014. PubMed ID: 24855258.

7-23. Clark G. A Farewell to Alms. Princeton University Press. 2007.

7-24. World Bank Group Databank, Description of Gini Index (https://databank.worldbank.org/metadataglossary/gender-statistics/series/SI.POV.GINI).

CHAPTER 8: PRESERVE KNOWLEDGE

8-1. Pettegree A, der Weduwen A. The Library. Basic Books. 2021.

8-2. Phillips H. Great Library of Alexandria. Social Science Open Access Repository. 2010. (https://nbn-resolving.org/urn:nbn:de:0168-ssoar-190807)

8-3. Hajar R. Past Glories: The Great Library of Alexandria. Heart Views. 1(7):278-282, 2000.

8-4. Dellavalle RP et al. Information science. Going, going, gone: lost Internet references. Science 302(5646):787-788, 2003. PubMed ID: 14593153.

8-5. Lawler A. Archaeology. Ten millennia of culture pilfered amid Baghdad chaos. Science 300(5618):402-403, 2003. PubMed ID: 12702844.

8-6. Eisenstein M. In pursuit of data immortality. Nature 604(7904):207-208, 2022. PubMed ID: 35379989.

8-7. Eve MP. Digital Scholarly Journals Are Poorly Preserved: A Study of 7 Million Articles. Journal of Librarianship and Scholarly Communication. 12(1): eP16288, 2024.

8-8. Hilton J. Lost Horizon. Pocket Books. 1933.

8-9. Miller WM Jr. A Canticle for Liebowitz. J. B. Lippincott & Co. 1959.

8-10. Niven L, Pournelle J. The Mote in God's Eye. Simon and Schuster. 1974.

8-11. Wilkinson MD et al. The FAIR Guiding Principles for scientific data management and stewardship. Scientific Data. 3:160018, 2016. PubMed ID: 26978244.

8-12. Gonzales S, Carson MB, Holmes K. Ten simple rules for maximizing the recommendations of the NIH data management and sharing plan. PLoS Computational Biology. 18(8):e1010397, 2022. PubMed ID: 35921268.

8-13. Johnson KR, Owens IFP; Global Collection Group. A global approach for natural history museum collections. Science. 379(6638):1192-1194, 2023. doi: 10.1126/science.adf6434. Epub 2023 Mar 23. PubMed ID: 36952410.

8-14. Vance TC, Huang T, Butler KA. Big data in Earth science: Emerging practice and promise. Science. 383(6688):eadh9607, 2024. PubMed ID: 38484080.

8-15. Crossref (https://www.crossref.org/).

8-16. CLOCKSS (https://clockss.org/).

8-17. LOCKSS (Lots of Copies Keep Stuff Safe) (https://www.lockss.org/).

8-18. DiSSCo (Distributed System of Scientific Collections) (https://www.dissco.eu/).

8-19. Digital Public Library of America (https://dp.la/).

8-20. Medina JJ et al. A rapid and cost-effective pipeline for digitization of museum specimens with 3D photogrammetry. PLoS One. 15(8):e0236417, 2020. PubMed ID: 32790700.

8-21. Ceze L, Nivala J, Strauss K. Molecular digital data storage using DNA. Nature Reviews Genetics. 20(8):456-466, 2019. PubMed ID: 31068682.

8-22. Doricchi A et al. Emerging Approaches to DNA Data Storage: Challenges and Prospects. ACS Nano. 16(11):17552-17571, 2022. PubMed ID: 36256971.

8-23. Chen F, Wu B. Laser-written glass tablets can preserve data for millennia. Nature. 650(8102):557-558, 2026. PubMed ID: 41709008.

8-24. Microsoft Research Project Silica Team. Laser writing in glass for dense, fast and efficient archival data storage. Nature. 650(8102):606-612, 2026. PubMed ID: 41709004.

8-25. Tanner A. Our Bodies, Our Data; How Companies Make Billions Selling Our Medical Records. Beacon Press. 2017.

8-26. Mandl KD, Perakslis ED. HIPAA and the Leak of "Deidentified" EHR Data. New England Journal of Medicine. 384(23):2171-2173, 2021. PubMed ID: 34110112.

8-27. Our World in Data, Costs of Information Storage. (https://ourworldindata.org/grapher/historical-cost-of-computer-memory-and-storage).

8-28. Rubin RE, Rubin RG. Foundations of Library and Information Science. Fifth Edition. Facet Publishing. 2020.

8-29. Susskind D. Growth, A History and a Reckoning. The Belknap Press of Harvard University Press. 2024.

CHAPTER 9: SET AND ACCOMPLISH LONG TERM GOALS

9-1. Visual Capitalist, the World's Top 50 Endowment Funds (https://www.visualcapitalist.com/worlds-top-endowment-funds/).

9-2. Sovereign Wealth Fund Institute, Top 100 Largest Endowments (https://www.swfinstitute.org/fund-rankings/endowment).

CHAPTER 10: CONCLUDING REMARKS

10-1. UNESCO Science Report (https://www.unesco.org/reports/science/2021/en) 2021.

10-2. Kim ES, Strecher VJ, Ryff CD. Purpose in life and use of preventive health care services. Proceedings of the National Academy of Sciences USA 111(46):16331-16336, 2014. PubMed ID: 25368165.

10-3. Alimujiang A et al. Association Between Life Purpose and Mortality Among US Adults Older Than 50 Years. JAMA Network Open 2(5):e194270, 2019. PubMed ID: 31125099.

10-4. Stavrova O. How much do sources of happiness vary across countries? A review of the empirical literature. Koelner Zeitschrift fuer Soziologie und Sozialpsychologie. 71(Suppl 1): 429-464, 2019.

10-5. Steptoe A. Happiness and Health. Annual Review of Public Health. 40:339-359, 2019. PubMed ID: 30601719.

10-6. Pinker S. The Better Anagels of our Nature, Why Violence Has Declined. Viking Press. 2011.

About the Author

JAMES WEBER IS A HUMAN geneticist. He received a PhD in biochemistry from the University of California-Berkeley in 1980. For roughly the first half of his professional career, he worked as the head of a large research laboratory. His discovery of short tandem repeat (microsatellite) DNA polymorphisms and construction of some of the first linkage maps of the human genome were important advances in human genetics. He is the author or coauthor of over 200 scientific journal articles and book chapters. During the second half of his career, Dr. Weber served as founder and CEO of one of the largest clinical DNA testing labs in the US. In this role, he oversaw the testing of about 500,000 patients.

More information about Dr. Weber can be found in Appendix 2 in this book, and on the Ten Generations Press web site (www.tengenerationspress.org).

www.ingramcontent.com/pod-product-compliance
Lightning Source LLC
LaVergne TN
LVHW100147061126
851857LV00010B/285
9798995825333